AIMMS

The User's Guide

AIMMS 3.7

October 2006

AIMMS

The User's Guide

Paragon Decision Technology

Johannes Bisschop
Marcel Roelofs

Paragon Decision Technology B.V.
Julianastraat 30
2012 ES Haarlem
The Netherlands
Tel.: +31(0)23-5511512
Fax: +31(0)23-5511517

Paragon Decision Technology Inc.
5400 Carillon Point
Kirkland, WA 98033
USA
Tel.: +1 425 576 4060
Fax: +1 425 576 4061

Email: info@aimms.com
WWW: www.aimms.com

ISBN 978-1-84753-782-9

AIMMS is a trademark of Paragon Decision Technology B.V. Other brands and their products are trademarks of their respective holders.

WINDOWS and EXCEL are registered trademarks of Microsoft Corporation. TEX, LATEX, and $\mathcal{AMS}$-LATEX are trademarks of the American Mathematical Society. LUCIDA is a registered trademark of Bigelow & Holmes Inc. ACROBAT is a registered trademark of Adobe Systems Inc.

Information in this document is subject to change without notice and does not represent a commitment on the part of Paragon Decision Technology B.V. The software described in this document is furnished under a license agreement and may only be used and copied in accordance with the terms of the agreement. The documentation may not, in whole or in part, be copied, photocopied, reproduced, translated, or reduced to any electronic medium or machine-readable form without prior consent, in writing, from Paragon Decision Technology B.V.

Paragon Decision Technology B.V. makes no representation or warranty with respect to the adequacy of this documentation or the programs which it describes for any particular purpose or with respect to its adequacy to produce any particular result. In no event shall Paragon Decision Technology B.V., its employees, its contractors or the authors of this documentation be liable for special, direct, indirect or consequential damages, losses, costs, charges, claims, demands, or claims for lost profits, fees or expenses of any nature or kind.

In addition to the foregoing, users should recognize that all complex software systems and their documentation contain errors and omissions. The authors, Paragon Decision Technology B.V. and its employees, and its contractors shall not be responsible under any circumstances for providing information or corrections to errors and omissions discovered at any time in this book or the software it describes, whether or not they are aware of the errors or omissions. The authors, Paragon Decision Technology B.V. and its employees, and its contractors do not recommend the use of the software described in this book for applications in which errors or omissions could threaten life, injury or significant loss.

This documentation was typeset by Paragon Decision Technology B.V. using LATEX and the LUCIDA font family.

About AIMMS

History

AIMMS was introduced by Paragon Decision Technology as a new type of mathematical modeling tool in 1993—an integrated combination of a modeling language, a graphical user interface, and numerical solvers. AIMMS has proven to be one of the world's most advanced development environments for building optimization-based decision support applications and advanced planning systems. Today, it is used by leading companies in a wide range of industries in areas such as supply chain management, energy management, production planning, logistics, forestry planning, and risk-, revenue-, and asset- management. In addition, AIMMS is used by universities worldwide for courses in Operations Research and Optimization Modeling, as well as for research and graduation projects.

What is AIMMS?

AIMMS is far more than just another mathematical modeling language. True, the modeling language is state of the art for sure, but alongside this, AIMMS offers a number of advanced modeling concepts not found in other languages, as well as a full graphical user interface both for developers and end-users. AIMMS includes world-class solvers (and solver links) for linear, mixed-integer, and nonlinear programming such as BARON, CPLEX, CONOPT, KNITRO, LGO, PATH, XA and XPRESS, and can be readily extended to incorporate other advanced commercial solvers available on the market today.

Mastering AIMMS

Mastering AIMMS is straightforward since the language concepts will be intuitive to Operations Research (OR) professionals, and the point-and-click graphical interface is easy to use. AIMMS comes with comprehensive documentation, available electronically and in book form.

Types of AIMMS applications

AIMMS provides an ideal platform for creating advanced prototypes that are then easily transformed into operational end-user systems. Such systems can than be used either as

- stand-alone applications, or
- optimization components.

Stand-alone applications Application developers and operations research experts use AIMMS to build complex and large scale optimization models and to create a graphical end-user interface around the model. AIMMS-based applications place the power of the most advanced mathematical modeling techniques directly into the hands of end-users, enabling them to rapidly improve the quality, service, profitability, and responsiveness of their operations.

Optimization components Independent Software Vendors and OEMs use AIMMS to create complex and large scale optimization components that complement their applications and web services developed in languages such as C++, Java, .NET, or Excel. Applications built with AIMMS-based optimization components have a shorter time-to-market, are more robust and are richer in features than would be possible through direct programming alone.

AIMMS *users* Companies using AIMMS include

- ABN AMRO
- Areva
- Bayer
- Bluescope Steel
- BP
- CST
- ExxonMobil
- Gaz de France
- Heineken
- Innovene
- Lufthansa
- Merck
- Owens Corning
- Perdigão
- Petrobras
- Philips
- PriceWaterhouseCoopers
- Reliance
- Repsol
- Shell
- Statoil
- Unilever

Universities using AIMMS include Budapest University of Technology, Carnegie Mellon University, George Mason University, Georgia Institute of Technology, Japan Advanced Institute of Science and Technology, London School of Economics, Nanyang Technological University, Rutgers University, Technical University of Eindhoven, Technische Universitt Berlin, UIC Bioengineering, Universidade Federal do Rio de Janeiro, University of Groningen, University of Pittsburgh, University of Warsaw, and University of the West of England.

A more detailed list of AIMMS users and reference cases can be found on our website www.aimms.com.

Contents

Preface

Three AIMMS
books

The printed AIMMS documentation consists of three books

- AIMMS—*The User's Guide*,
- AIMMS—*The Language Reference*, and
- AIMMS—*Optimization Modeling*.

The first two books emphasize different aspects in the use of the AIMMS system, while the third book is a general introduction to optimization modeling. All books can be used independently.

Available online

In addition to the printed versions, these books are also available on-line in the ADOBE Portable Document Format (PDF). Although new printed versions of the documentation will become available with every new functional AIMMS release, small additions to the system and small changes in its functionality in between functional releases are always directly reflected in the online documentation, but not necessarily in the printed material. Therefore, the online versions of the AIMMS books that come with a particular version of the system should be considered as the authoritative documentation describing the functionality regarding that particular AIMMS version.

Release notes

Which changes and bug fixes are included in particular AIMMS releases are described in the associated release notes.

What is new in AIMMS 3.7

*What is new in
AIMMS 3.7?*

This documentation reflects the state of AIMMS version 3.7. Compared to AIMMS 3.6, the following major new and extended functionalities have been added to the system:

- integrated GIS support in the AIMMS network flow object,
- syntax editor,
- stochastic programming support,
- parallel solver sessions,
- improved multi-developer support,
- extensions to the pivot table object,
- extensions to the web services support,

- solver additions and updates, and
- additional supported platforms.

Integrated GIS support

The network flow object of AIMMS 3.7 has been extended with the capability to dynamically retrieve background images from one or more GIS servers, as illustrated in Figure 1. The GIS-generated background can consist of several layers, possibly obtained from multiple GIS servers. Alternatively, GIS files can be generated from within your AIMMS application. AIMMS supports retrieval of data layers from any GIS server that supports the WMS, WFS and GML standards.

Integrated GIS support

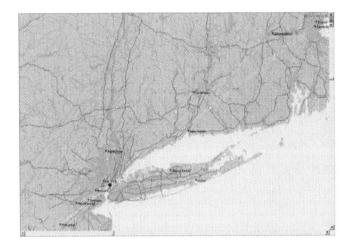

Figure 1: Example of a network flow object with a GIS-generated background

This GIS support of AIMMS will allow end-users to depict their geographical related solution data on a dynamic map which will enhance visualization, and ease the interpretation and modification of input/solution data for both end-users and developers within our integrated AIMMS environment.

Benefits

Syntax editor

AIMMS 3.7 has been extended with a state-of-the-art syntax highlighting control as illustrated in Figure 2, with support for such features as:

Syntax editor

- user-configurable formatting for keywords, elements, strings, and identifier types,
- collapsible code blocks,
- bracket matching capabilities,

- auto-indentation and indentation guidelines,
- member lists (e.g. to select an identifier from a namespace, or a suffix), and
- function and procedure prototype tooltips.

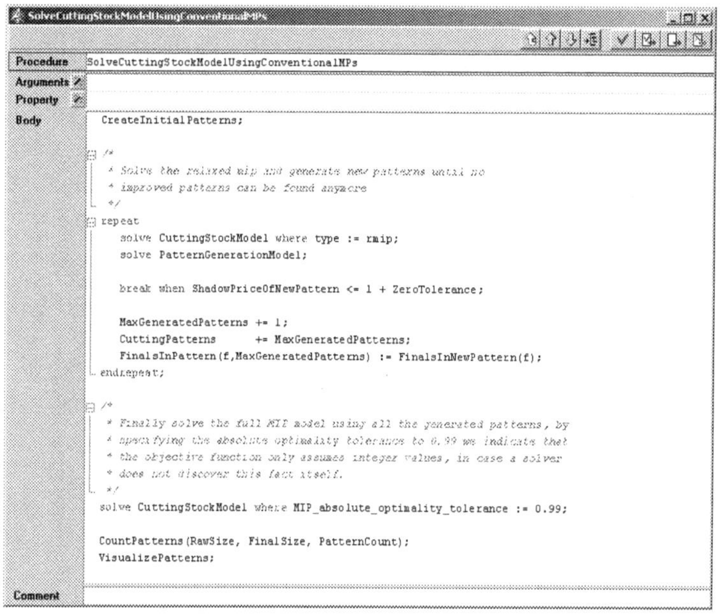

Figure 2: Example of AIMMS syntax editor

Benefits   With the introduction of a syntax highlighting control in AIMMS, modifying AIMMS source code and visual code inspection will become much easier. Also, the user configurable formatting (personal profiles) and coloring makes it possible to clearly and visibly distinguish between variables and parameters in, for instance, constraint definitions.

Stochastic programming support

Stochastic programming support   AIMMS 3.7 will offer support for generating a stochastic LP/MIP recourse model from any given deterministic model, without the need to reformulate the deterministic model. By only supplying additional attributes for selected parameters, variables and constraints, AIMMS can generate both a deterministic and recourse model from the same formulation. Stochastic programming in AIMMS is discussed in full detail in Chapter 18 of the Language Reference.

To solve the recourse model, AIMMS 3.7 will generate and solve the corresponding deterministic equivalent, either with explicitly or implicitly added non-anticipativity constraints. Support for an open stochastic Benders solution scheme (implemented using the GMP library) is planned for AIMMS 3.8.

Solving stochastic models

Various user adaptable templates for generating a scenario tree and the corresponding stochastic input data for the recourse model are available in the form of a system module. This module can be imported into any AIMMS model implementing a stochastic model.

Scenario generation

The support of stochastic programming within AIMMS allows one to solve mathematical models with uncertainty to optimality and create robust solutions without structural changes to the underlying deterministic model.

Benefits

Parallel solver sessions

The GMP library of AIMMS 3.7 has been extended with support for parallel solver sessions, i.e. solver session running asynchronously in a separate process or thread of execution. This allows a modeling application to solve multiple math programs in parallel on one or more processors. Support for parallel solver sessions is discussed in more detail in Chapter 19 of the Language Reference.

Parallel solver sessions

The controlling AIMMS model will be notified when an asynchronous solver session has completed, or when some user-definable event has occurred. This allows an AIMMS developer to create advanced solution algorithms for demanding applications that use the available computer resources to their full capacity.

Synchronization

For advanced optimization applications in which multiple independent mathematical programs can be solved simultaneously, this may dramatically increase the application performance on multi-processor computers or multi-core processors.

Benefits

Improved multi-developer support

AIMMS 3.7 offers substantially improved support for multiple developers to work on a single AIMMS application. AIMMS 3.7 applications can be divided into multiple library projects for relatively independent tasks that can be distinguished in the application, where the main project should glue together the parts of the model and GUI provided by the individual library projects. Library projects are discussed in more detail in Chapter 3 of the User's Guide.

Improved multi-developer support

Comparison with modules

Compared to the source code modules supported in AIMMS 3.2 and up, library projects can not only contain a part of the applications model source, but also the associated pages, templates and menus that must be included in the overall application end-user GUI on behalf of the library. Library modules provide an interface of public identifiers, while keeping all identifiers not in the interface strictly private. This allows library developers to work independently on their assigned sub-projects.

Benefits

This feature will significantly increase AIMMS capabilities to allow multiple developers to work on all aspects (i.e. model and GUI) of a single application. It also allows for effective reuse of shareable parts of an AIMMS application.

Pivot table extensions

Pivot table extensions

AIMMS 3.7 offers various extensions to the graphical pivot table object introduced in AIMMS 3.6, including:

- support for domain identifiers to influence the sparsity of a pivot table,
- advanced sorting capabilities of rows and/or columns,
- allow identifier suffices to count as an extra dimension,
- support for check boxes as cell values,
- support for grand totals, and
- ability to extend the table with several layers of additional indices that can be mapped to existing indices.

The latter extension enables application-specific aggregations. For instance, adding a set of quarters that are mapped to months and a set of years that are mapped to quarters, allows the display of both quarterly and yearly totals for data that is defined over months only.

Benefits

Even more functionality and flexibility is available to provide more power to both the developer and end user to visualize and inspect data in the pivot table objects.

Web services extensions

Web services extensions

The support for web services introduced in AIMMS 3.6 has been significantly extended in AIMMS 3.7. Web services in AIMMS 3.7 support full asynchronous messaging between a community of AIMMS agents, and any external application that implements a web service according to an AIMMS-generated WSDL description of the service.

The web service listener process of AIMMS has been implemented as a separate service in AIMMS 3.7. This has significantly eased the configuration of AIMMS web services compared to AIMMS 3.6, where the listener process was implemented as a rather hard to configure virtual directory under the IIS web server.

Easier configuration

The extended set-up allows for a much more effective use of AIMMS in Service Oriented Architectures (SOA). In addition, the new support for asynchronous messaging allows the creation of a distributed application consisting of one or more AIMMS instances and external applications.

Benefits

Solver additions and updates

AIMMS 3.7 has been extended with a link to the KNITRO 5.0 solver from Ziena Optimization, Inc. KNITRO is a world-class solver for NLP, complementarity and MPCC (mathematical programs with complementarity constraints, also commonly known as MPEC) models. The KNITRO solver can be purchased as an add-on to your AIMMS system.

KNITRO *5.0*

AIMMS 3.7 supports the MPCC class by allowing you to formulate optimization problems with complementarity constraints in their constraint sets. To solve MPCC models, AIMMS requires a link to the KNITRO solver. Solving MPCC models is discussed in full detail in Section 17.4 of the Language Reference.

Support in AIMMS

The CPLEX solver has been updated to version 10.1. Compared to CPLEX 9, CPLEX 10 provides

CPLEX *10*

- improved performance,
- improved infeasibility analysis,
- solution polishing for finding the best solution for difficult MIP problems,
- advanced restart capabilities for MIP problems, and
- indicator constraints, a very efficient way of controlling whether or not a constraint takes effect, based on the value of a binary variable.

To support the indicator constraints feature of CPLEX 10, constraint declarations in AIMMS now support a new property IndicatorConstraint, which activates a new attribute ACTIVATING CONDITION. The value of this attribute is directly translated to the CPLEX 10 indicator constraints facility. The support for indicator constraints in AIMMS is discussed in full detail in Section 14.2.3 of the Language Reference.

Support in AIMMS

BARON *7.5*

The link in AIMMS 3.7 to the BARON global optimization solver has been updated to the latest version 7.5. This version of BARON supports two new types of constraints, convex and relaxation-only constraints. Through these new constraint types you can manually pass particular knowledge to BARON about your model which cannot be automatically detected by the BARON solver itself. Having this knowledge available explicitly may enable the BARON solver to speed up its solution process dramatically.

Support in
AIMMS

AIMMS 3.7 supports these new constraint types for the BARON solver through the .Convex and .RelaxationOnly suffices for constraints in your model. The use of these suffices is explained in full detail in Section 14.2.5 of the Language Reference.

Benefits

By adding new solvers, and keeping up-to-date with existing solvers and solver features, AIMMS allows you to

■ solve more model types, and
■ solve existing models more efficiently.

Supported platforms

Linux x86_64 port

AIMMS 3.7 will be available for Linux x86_64 RHEL 4 platform. This port supports the same functionality as the Linux x86 version, while the 64-bit addressing allows larger models to be solved.

Support for Windows Vista

AIMMS 3.7 has been tested on the latest available beta versions of Windows Vista, and will be supported on Windows Vista as soon as it is officially released by Microsoft.

Benefits

By extending the support for AIMMS to more computing platforms and updating the support for existing platforms, you are able to benefit from improved hardware and operating system architectures.

What is in the AIMMS documentation

The User's Guide

The AIMMS User's Guide provides a global overview of how to use the AIMMS system itself. It is aimed at application builders, and explores AIMMS' capabilities to help you create a model-based application in a easy and maintainable manner. The guide describes the various graphical tools that the AIMMS system offers for this task. It is divided into five parts.

■ Part I—*Introduction to* AIMMS—what is AIMMS and how to use it.
■ Part II—*Creating and Managing a Model*—how to create a new model in AIMMS or manage an existing model.

- Part III—*Creating an End-User Interface*—how to create an intuitive and interactive end-user interface around a working model formulation.
- Part IV—*Data Management*—how to work with cases and datasets.
- Part V—*Miscellaneous*—various other aspects of AIMMS which may be relevant when creating a model-based end-user application.

The AIMMS Language Reference provides a complete description of the AIMMS modeling language, its underlying data structures and advanced language constructs. It is aimed at model builders only, and provides the ultimate reference to the model constructs that you can use to get the most out of your model formulations. The guide is divided into seven parts.

The Language Reference

- Part I—*Preliminaries*—provides an introduction to, and overview of, the basic language concepts.
- Part II—*Nonprocedural Language Components*—describes AIMMS' basic data types, expressions, and evaluation structures.
- Part III—*Procedural Language Components*—describes AIMMS' capabilities to implement customized algorithms using various execution and flow control statements, as well as internal and external procedures and functions.
- Part IV—*Sparse Execution*—describes the fine details of the sparse execution engine underlying the AIMMS system.
- Part V—*Optimization Modeling Components*—describes the concepts of variables, constraints and mathematical programs required to specify an optimization model.
- Part VI—*Data Communication Components*—how to import and export data from various data sources, and create customized reports.
- Part VII—*Advanced Language Components*—describes various advanced language features, such as the use of units, modeling of time and communicating with the end-user.

The book on optimization modeling provides not only an introduction to modeling but also a suite of worked examples. It is aimed at users who are new to modeling and those who have limited modeling experience. Both basic concepts and more advanced modeling techniques are discussed. The book is divided into five parts:

Optimization Modeling

- Part I—*Introduction to Optimization Modeling*—covers what models are, where they come from, and how they are used.
- Part II—*General Optimization Modeling Tricks*—includes mathematical concepts and general modeling techniques.
- Part III—*Basic Optimization Modeling Applications*—builds on an understanding of general modeling principles and provides introductory application-specific examples of models and the modeling process.
- Part IV—*Intermediate Optimization Modeling Applications*—is similar to part III, but with examples that require more effort and analysis to con-

struct the corresponding models.

- Part V—*Advanced Optimization Modeling Applications*—provides applications where mathematical concepts are required for the formulation and solution of the underlying models.

Documentation of deployment features

In addition to the three major AIMMS books, there are several separate documents describing various deployment features of the AIMMS software. They are:

- AIMMS—*The Function Reference,*
- AIMMS—*The COM Object User's Guide and Reference,*
- AIMMS—*The Multi Agent and Web Services User's Guide,*
- AIMMS—*The Excel Add-In User's Guide,* and
- AIMMS—*The Open Solver Interface User's Guide and Reference.*

These documents are only available in PDF format.

Help files

The AIMMS documentation is complemented with a number of help files that discuss the finer details of particular aspects of the AIMMS system. Help files are available to describe:

- the execution and solver options which you can set to globally influence the behavior of the AIMMS' execution engine,
- the finer details of working with the graphical modeling tools, and
- a complete description of the properties of end-user screens and the graphical data objects which you can use to influence the behavior and appearance of an end-user interface built around your model.

The AIMMS help files are both available as Windows help files, as well as in PDF format.

AIMMS *tutorials*

Two tutorials on AIMMS in PDF format provide you with some initial working knowledge of the system and its language. One tutorial is intended for beginning users, while the other is aimed at professional users of AIMMS.

Searching the documentation

As the entire AIMMS documentation is available in PDF format, you can use the search functionality of Acrobat Reader to search through all AIMMS documentation for the information you are looking for. From within the **Help** menu of the AIMMS software you can access a pre-built search index to quicken the search process.

AIMMS *model library*

AIMMS comes with an extensive model library, which contains a variety of examples to illustrate simple and advanced applications containing particular aspects of both the language and the graphical user interface. You can find the AIMMS model library in the Examples directory in the AIMMS installation directory. The Examples directory also contains an AIMMS project providing an index to all examples, which you can use to search for examples that illustrate specific aspects of AIMMS.

What is in the User's Guide

Part I of the User's Guide provides a basic introduction to AIMMS, its position among other technologies, and its use.

Introduction to AIMMS

- Chapter 1—AIMMS *and Analytic Decision Support*—discusses the concept of Analytic Decision Support (ADS), AIMMS as an ADS development environment, as well as a comparison to other ADS tools.
- Chapter 2—*Getting Started*—explains how to create a new AIMMS application, and provides an overview of both the modeling tools available in AIMMS and the files associated with an AIMMS project.
- Chapter 3—*Organizing a Project into Libraries*—describes the facilities available in AIMMS to allow multiple developers to collaborate on a single project.

Part II discusses all aspects of the AIMMS system that are relevant for entering and maintaining the model source associated with a particular modeling application.

Creating and managing a model

- Chapter 4—*The Model Explorer*—introduces the main graphical tool available in AIMMS for accessing the model source. It discusses various aspects specific to the model tree, as well as the basic concepts common to all trees used in the AIMMS system.
- Chapter 5—*Identifier Declarations*—explains how you can add identifier declarations to the model tree, and how you can modify the various attributes of an identifier in its attribute window.
- Chapter 6—*Procedures and Functions*—explains how you can create procedures and functions within your model, how to add arguments to such procedures and functions, and describes the AIMMS concepts that help you to sub-divide procedure and function bodies into smaller more meaningful entities.
- Chapter 7—*Viewing Identifier Selections*—discusses the flexible identifier selector tool in AIMMS, which allows you to create and simultaneously view selections of identifiers in your model.
- Chapter 8—*Debugging and Profiling an* AIMMS *Model*—discusses AIMMS' debugger and profiler, which can help you to track the modeling errors in an AIMMS model, or to find and speed up time-consuming statements in your model.
- Chapter 9—*The Math Program Inspector*—introduces a graphical debugging tool for finding infeasibilities or unexpected results of a math program contained in your model.

Creating an end-user interface

Part III introduces the fundamental concepts and design tools available in AIMMS to create a graphical end-user interface for your modeling application, as well as AIMMS' reporting facilities.

- Chapter 10—*Pages and Page Objects*—introduces the AIMMS concept of end-user pages. In addition, it explains how to add graphical (data) objects to such pages, and how to link these data objects to identifiers in your model.
- Chapter 11—*Page and Page Object Properties*—discusses the options for pages and page objects that you can modify to alter the behavior and appearance of your end-user interface.
- Chapter 12—*Page Management Tools*—describes the AIMMS tools that can help you create and manage a large collection of end-user pages in an easily maintainable fashion.
- Chapter 13—*Page Resizability*—explains the basic concepts available in AIMMS to define the behavior of pages when resizing.
- Chapter 14—*Creating Printed Reports*—discusses the concept of print pages which you can use to create a printed report of your model results.
- Chapter 15—*Designing End-User Interfaces*—provides some background on designing professional end-user interfaces that are both easy to use and to maintain.

Data management

Part IV focuses on the facilities within AIMMS for performing common and advanced case management tasks.

- Chapter 16—*Case Management*—describes the basic case management facilities and tools in AIMMS. It also provides you with an overview of AIMMS' capabilities to start a batch run of a model with multiple cases, and to work with multiple case data, both in the model and the end-user interface.
- Chapter 17—*Advanced Data Management*—explains the advanced concepts of case types, data categories and datasets, and their interaction with cases. In addition, this chapter discusses the security aspects of cases and datasets, as well as AIMMS' capabilities to access the tree of all cases and datasets from within the modeling language.

Miscellaneous

Part V discusses the various miscellaneous concepts that may be of interest to both AIMMS developers and/or end-users.

- Chapter 18—*User Interface Language Components*—provides a complete overview of the function library available in AIMMS for communication with the end-user through the various tools available in the AIMMS end-user interface.
- Chapter 19—*Calling* AIMMS—describes AIMMS' command line options, the restrictions with respect to end-user licenses, and the possibilities of calling an AIMMS model from within your own application.

- Chapter 20—*Project Security*—discusses various security aspects such as protecting your project through a VAR license, adding a user database to a model to provide user authentication, and case file security.
- Chapter 21—*Project Settings and Options*—describes the tools available in AIMMS to alter the execution behavior of your model, the appearance of its interface, and various other aspects concerning AIMMS itself and its solvers.
- Chapter 22—*Localization and Unicode Support*—discusses AIMMS' built-in support for localizing the end-user interface of your project (i.e. making it capable of dealing with multiple languages). It also discusses the capabilities and limitations of the AIMMS Unicode version, which, combined with the localization features, allows you to create AIMMS end-user applications for the Asian market.

The authors

Johannes Bisschop received his Ph.D. in Mathematical Sciences from the Johns Hopkins University in Baltimore USA in 1974. From 1975 to 1980 he worked as a Researcher in the Development Research Center of the World Bank in Washington DC, USA. In 1980 he returned to The Netherlands and accepted a position as a Research Mathematician at Shell Research in Amsterdam. After some years he also accepted a second part-time position as a full professor in the Applied Mathematics Department at the Technical University of Twente. From 1989 to 2003 he combined his part-time position at the University with managing Paragon Decision Technology B.V. and the continuing development of AIMMS. From 2003 to 2005 he held the position of president of Paragon Decision Technology B.V. His main interests are in the areas of computational optimization and modeling.

Johannes Bisschop

Marcel Roelofs received his Ph.D. in Applied Mathematics from the Technical University of Twente in 1993 on the application of Computer Algebra in Mathematical Physics. From 1993 to 1995 he worked as a post-doc at the Centre for Mathematics and Computer Science (CWI) in Amsterdam in the area of Computer Algebra, and had a part-time position at the Research Institute for the Application of Computer Algebra. In 1995 he accepted his current position as CTO of Paragon Decision Technology B.V. His main responsibilities are the design and documentation of the AIMMS language and user interface.

Marcel Roelofs

In addition to the main authors, various current and former employees of Paragon Decision Technology B.V. and external consultants have made a contribution to the AIMMS documentation. They are (in alphabetical order):

Other contributors to AIMMS

- Pim Beers
- John Boers
- Peter Bonsma
- Mischa Bronstring
- Ximena Cerda Salzmann
- Michelle Chamalaun
- Robert Entriken
- Thorsten Gragert
- Koos Heerink
- Nico van den Hijligenberg
- Marcel Hunting
- Roel Janssen

- Gertjan Kloosterman
- Joris Koster
- Chris Kuip
- Gertjan de Lange
- Ovidiu Listes
- Bianca Makkink
- Peter Nieuwesteeg
- Giles Stacey
- Richard Stegeman
- Selvy Suwanto
- Jacques de Swart
- Martine Uyterlinde

Part I

Introduction to AIMMS

Chapter 1

AIMMS and Analytic Decision Support

The acronym AIMMS stands for

Advanced Integrated Multidimensional Modeling Software.

What is AIMMS?

AIMMS offers you an easy-to-use and all-round development environment for creating fully functional *Analytic Decision Support* (ADS) applications ready for use by end-users. The software is constructed to run in different modes to support two primary user groups: *modelers* (application developers) and *end-users* (decision makers). AIMMS provides the ability to place all of the power of the most advanced mathematical modeling techniques directly into the hands of the people who need this to make decisions.

This chapter is aimed at first-time users of the AIMMS modeling system. In a nutshell, it provides

This chapter

- a description of the characteristics of Analytic Decision Support (ADS) applications,
- an overview of AIMMS as an ADS development environment, and
- some examples of its use in real-life applications.

1.1 Analytic decision support

Analytic decision support applications are usually interactive decision support systems with a strong internal emphasis on *advanced computational techniques* and that pertain to extensive problem analysis on the outside. They typically

Analytic decision support

- represent a complex and large-scale reality,
- organize and use large amounts of interrelated multidimensional data based on corporate and market information,
- use advanced arithmetic manipulations and/or optimization tools to find a solution,
- apply analytic techniques or perform "what-if" experiments to assess the consequences of making a decision under different scenarios,
- employ advanced visualization techniques to provide an insight into the solution and/or the problem complexity, and

- are subject to permanent change due to a changing reality or improved insights.

*Increasing
market need*

With the world becoming daily more complex, decision makers around the world are in search of advanced decision support tools. Such tools can help them get insights into their decision problems, monitor the consequences of previous decisions, and help them take new decisions on a regular basis. There is substantial evidence that analytic decision support applications are becoming increasingly popular throughout industry and government, as the improved decisions generated by ADS applications imply increased profit and/or efficiency.

*Supporting
developments*

A number of major developments in the last decade have increased the suitability of analytic decision support to tackle such problems:

- corporate databases are becoming increasingly mature and allow a quick follow-up to market changes,
- the increasing speed of PCs allows interactive use, even with complex applications,
- the visually attractive and convenient presentation using the standardized and user-friendly Windows environment makes complex processes more accessible to decision makers, and
- the availability of standardized and improved optimization tools allows ADS application developers to specify the problem without having to specify a complicated algorithm to solve it.

*Applicable
problem areas*

Analytic decision support lends itself to a wide variety of decision support problems. The following list provides a non-exhaustive overview of the areas in which analytic decision support is applicable:

- strategic and tactical planning of resources in industry and government,
- operational scheduling of machines, vehicles, product flow and personnel,
- strategic evaluation studies in the areas of energy, environment, forestry and social policies,
- financial decision-making to support asset-liability management,
- economic decision-making to control market clearing and economic development, and
- technical decision-making to support the design and calibration of systems and objects.

1.2 AIMMS as an ADS development environment

As an ADS development environment, AIMMS possesses a unique combination of advanced features and design tools which allow you to build complex
ADS applications which are easily maintainable—in a fraction of the time required with conventional programming languages. Figure 1.1 provides a top-
level overview of the components available in AIMMS.

*AIMMS as ADS
development
environment*

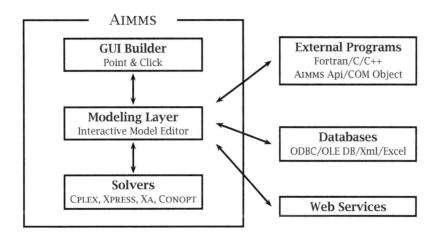

Figure 1.1: Graphical overview of AIMMS components

The multidimensional modeling language in AIMMS offers a powerful index
notation which enables you to capture the complexity of real-world problems
in an intuitive manner. In addition, the language allows you to express very
complex relationships in a compact manner without the need to worry about
memory management or sparse data storage considerations. The combined
declarations and procedures using these multidimensional structures can be
organized, edited and displayed using an advanced interactive model editor.

*Multidimen-
sional modeling
language*

One of the outstanding features of AIMMS is the capability of specifying and
solving linear and nonlinear constraint-based optimization models. Using the
same compact and rich notation available for procedural statements, symbolic
constraints can be formulated in a simple and concise manner. With only a
single instruction, an optimization model can be transferred to, and solved by,
world- class solvers such as CPLEX, XPRESS, XA and CONOPT.

*Optimization
modeling*

Advanced language features

Selected advanced Aimms language features include:

- a rich set of mathematical, statistical and financial functions,
- a powerful combination of (automatically updated) multidimensional definitions and procedural execution,
- the ability to easily express time-based models through the use of calendars and horizons, including support for rolling horizons with automatic aggregation and disaggregation, and
- the ability to associate units of measurement with model identifiers assuring unit consistency within expressions.

Integrated GUI builder

In addition to its versatile modeling language Aimms offers an integrated tool for constructing a custom graphical user interface (GUI) for your decision support application. End-user screens can be created in an easy point-and-click manner, and can include such common graphical objects as tables, charts and curves, all closely linked to multidimensional identifiers in your model. Included, amongst other more advanced objects, are a Gantt chart for visualizing time-phased planning/scheduling applications, and a network flow object for visualizing two-dimensional maps and flows.

Advanced GUI tools

To support you in creating complete end-user interfaces in a quick and maintainable fashion, Aimms offers the following advanced tools:

- the *template manager* enables you to create a uniform look and feel by allowing common objects (such as headers, footers, and navigation buttons) to be placed on hierarchically organized templates which can be inherited by multiple pages,
- the *page manager* allows you to specify a natural page order, with which you can guide an end-user through your application by adding special page manager-driven navigation controls to templates or pages,
- the *menu builder* enables you to create customized end-user menus and toolbars to be added to your templates and pages,
- the *identifier selection wizard* assists you not only in selecting complete model identifiers, or slices thereof, for graphical display, but also in quickly linking data from various page objects.

Integrated case management

Case management forms an important part of any decision support application, and enables end-users to run the model with varying scenarios. Aimms also offers advanced data management, which allows you to create data categories for holding blocks of related data (for instance topology data, or supply and demand scenarios). Data sets associated with these data categories can be combined to form a single case, and thus can be shared by more than one case. In addition, to perform an extensive what-if analysis, you can select a large number of cases and run them in batch mode overnight.

As data form the life blood of any decision support application, Aimms offers extensive facilities to link your application to corporate databases using ODBC or OLE DB. Specialized wizards help you relate columns in a database table with the corresponding multidimensional identifiers in your Aimms model. Once you have established such relationships, you can specify straightforward read and write statements to transfer data to and from the database.

Database connectivity

To facilitate the re-use of existing code, or to speed up computationally intensive parts of your application, Aimms allows you to execute external procedures or functions in a DLL from within your model. External functions can even be used within the constraints of an optimization model. In addition, Aimms offers an Application Programming Interface (API) as well as a number of COM interfaces which enables you to use your Aimms model as a component from within an external application, to communicate data in a sparse fashion, and to execute procedures written in Aimms.

Linkages to other applications

Any Aimms project can be configured to be exposed as a web service. This means that you can invoke a set of Aimms procedures in your project from anywhere on the Internet, provided that you have a client application that features the calling of web services by the use of SOAP requests. Your client application does not necessarily have to adhere to the default Aimms XML data format: by using so-called *attachment arguments* in your procedures, you can send your data in the format of your choice, provided that a mapping can be specified between this data and the identifiers in your Aimms project.

Web services

The Aimms system has integrated facilities to create a database of end-users and link this database to one or more Aimms-based applications. The end-user database contains information on the level of authorization of all end-users within an application. Through these authorization levels you can specify whether an end-user is allowed to access case data, view pages, modify data, and execute particular parts of the model.

User management

The development of a professional decision support application usually represents a considerable investment in time and thus money. Aimms offers facilities to protect this investment and to prevent unauthorized use of particular applications. Your project and the source code of your model can be shielded by using a security scheme based upon your own unique customer code. In addition, Aimms allows you to create your own application-specific VAR licenses to restrict either the number of (concurrent) users or the lifespan of a license.

Protecting your investment

Extensive documentation

AIMMS comes complete with extensive documentation in the form of three books:

- a User's Guide to explain the overall functionality and use of AIMMS,
- a Language Reference giving a detailed description of the AIMMS language, and
- a Modeling Guide introducing you to both basic and advanced modeling techniques.

All of these books are available in hard copy as well as in electronic form. In addition, each system comes complete with a collection of example applications elucidating particular aspects of the language and end-user interface.

1.3 What is AIMMS used for?

AIMMS usage

AIMMS is used worldwide as a development environment for all kinds of analytic decision support applications. To give you some insight into the areas in which AIMMS has been used successfully, this section describes a small subset of ADS applications, namely those in which Paragon Decision Technology itself has been involved (sometimes actively, sometimes at a distance).

Crude oil scheduling

The crude oil scheduling system covers the allocation, timetabling, blending and sequencing activities from the waterfront (arrival of crude ships) via the crude pipeline to the crude distillation units. The result is a schedule for the discharge of crudes, the pipeline and the crude distillers (sequencing, timing and sizing of crude batches), plus planning indications on the arrival of new crude deliveries. Enhanced decision support includes improved and timely reaction to changes and opportunities (e.g. distressed crude cargoes, ship and pumping delays, operation disturbances) and improved integration between crude acquisition and unit scheduling.

Strategic forest management

The strategic forest management system allows foresters to interactively represent large forested areas at a strategic level. Such a flexible decision framework can help in understanding how a forest develops over time. The system also allows one to explore forest management objectives and their trade-offs, plus the preparation of long-term forest management plans.

Transport scheduling in breweries

The transport scheduling system for breweries allows end-users to interactively steer the flow of products through all phases of the brewing process from hops to bottled beer. The application can be used either in an automatic mode where the flow of products is totally determined by the system, or it can be used in a manual mode where the user can set or alter the flow using the Gantt chart. The system can also be used in a simulation mode to test the response of the entire brewery to varying demands over a longer period of time.

The risk management system for market makers and option traders has a wide functionality including the theoretical evaluation of derivatives, an extensive sensitivity analysis, the display of risk profiles, the generation of scenarios, the generation of price quotes and exercise signals, minimization of risk exposure, the calculation of exercise premiums and implied data (volatilities and interest rates), plus an overview of all transactions for any day.

Risk management

The refinery blending system is a blend scheduling and mixture optimization system. It is able to handle the complete pooling and blending problem, and optimizes both the blend schedules and the mixes under a complete set of (real-life) operational constraints. The system offers user flexibility in that the user can decide upon the number of components, fuel mixtures, long versus short term scheduling, and stand-alone versus refinery-wide scheduling.

Refinery blending

Catalytic cracking refers to a refining process in which hydrocarbons are converted into products with a lower molecular mass. The catalytic cracking support system has three major components: (a) a graphical user interface consisting of interactive pages, validation routines, plus reporting and data handling facilities, (b) the model equations, including those for heat, yields, product properties, economics, and (c) an on-line process control environment with an off-line mode in which multiple studies with differing parameters and variables can be compared.

Catalytic cracker support

Data reconciliation is the process of making the smallest possible adjustment to a collection of measurements within a system such that the adjusted data values satisfy all the balance constraints applicable to the system. In the particular application in question, data reconciliation was applied to a chemical process, requiring that the relevant mass, component and thermodynamic balances be satisfied for all units within the system.

Data reconciliation

1.4 Comparison with other ADS tools

There are several tools available in the market that can, in principle, be used as a development environment for analytic decision support applications. The most well-known are:

ADS development tools

- spreadsheets,
- databases,
- programming languages, and
- multidimensional modeling languages.

Comparison   Spreadsheets, databases and programming languages all have their strengths as development tools for a large variety of applications. Advanced modeling systems such as Aimms should not be seen as a complete replacement for these three development environments, but rather as a tool specifically designed for developing analytic decision support applications. The following paragraphs outline the advantages and disadvantages of each of these tools with respect to their suitability as a development environment for ADS.

Spreadsheet   If you are a fervent spreadsheet user, it seems only natural to build your ADS applications on top of a spreadsheet. However, this may not always be the best choice. A spreadsheet approach works well when:

- you don't need to specify a large number of relationships,
- there are only a few procedures to be written,
- the size of your data sets remains stable,
- the need to add or remove dimensions is limited, and
- you will carry out all the maintenance activities yourself.

When this is not the case, the Aimms approach may offer a suitable alternative, because:

- specifying a large number of (often similar) relationships can be done using indexed identifiers and definitions for these identifiers,
- adding and managing both internal and external procedures is a straightforward task using the Aimms language and model editor,
- modifying the size of any (index) set in Aimms is natural, as there is a complete separation between structure and data,
- adding or removing dimensions takes place in the language and does not require the copying of cells or creating more worksheets, and
- not only can the structure of the entire model be made visible, but also the model editor allows someone else to create customized overviews of model structure for further maintenance.

Database   If you are a fervent database user, it seems only natural to build your ADS applications using a language such as Visual-C/C++, Delphi or PowerBuilder on top of a database such as Microsoft Access, and Oracle. However, this may not always be the best choice. Using a database approach works well when:

- all of the data for your application is already stored in a database,
- the end-user GUI requires relatively little programming,
- speed of data transfer is not crucial,
- there is a limited need to link to external solvers, and
- maintenance is carried out by yourself or another experienced programmer.

When this is not the case, the AIMMS approach may offer a suitable alternative, because:

- data transfer works well not only for data stored in a database, but also for data in ASCII and case files,
- the compact modeling language combined with the point-and-click GUI builder minimizes the amount of programming required,
- internal data transfer during (the sparse) execution is extremely fast and does not require the repeated transfer of data between external programs,
- standard links to solvers are built into AIMMS, and
- compact and simple data structures on the one hand, and point-and-click GUI construction on the other hand, help ease maintenance.

If you are a fervent programmer, it seems only natural to build your ADS applications using languages such as C/C++ or Fortran. However, this may not always be the best choice. Using a programming language works well when:

Programming language

- efficient data structures require relatively little effort,
- there are many procedures to be written,
- development times are not crucial,
- there is a limited need to link to external programs, and
- maintenance is carried out by yourself or another experienced programmer.

When this is not the case, the AIMMS approach may offer a suitable alternative, because:

- the standard multidimensional data structures in AIMMS require no special effort, and are efficient since all data storage and data manipulations are carried out in a sparse manner,
- writing procedures in AIMMS is at least as easy as in a programming language: their control structures are similar, and AIMMS has the advantage that no special data structures are required,
- specially developed tools for the construction of programs and GUIs minimize development time,
- standard links to databases and solvers are built into AIMMS, and
- compact and simple data structures on the one hand, and point-and-click GUI construction on the other, help to ease maintenance.

Comparison summary

Table 1.1 summarizes the most important issues that determine the suitability of the above development tools as a development environment for ADS applications. The table focuses on

- the initial development effort to create an ADS application,
- the subsequent time required for product maintenance (extremely important due to the permanently changing nature of ADS applications), and
- the openness of the environment with respect to data entry formats and third party components.

A '+' indicates that the product scores well in this area, a '-' indicates that it does not perform well in this area.

Building tool	Development time	Maintenance time	Openness	Suitability as an ADS tool
Spreadsheet	+	--	++	+
Database	+	-	+	+
Programming language	-	-	++	++
AImms	++	++	+	++

Table 1.1: Comparison of ADS development tools

Developer quote

In support of the comparison in Table 1.1, the following quote, from one of our customers, clearly expresses the advantages of using AImms as a development environment for ADS applications.

"Software development requires four tasks: definition, design, implementation and testing. When using AImms, the focus is on definition. The result is an implementation which can be immediately tested. I now spend the majority of my time working on the customer's problem, and verifying that we have got the requirements correct. My job is now that of an applications engineer, rather than a software engineer. One of our customers stated that our recent project with them (using AImms) was the first software project in their history not to have a single 'Software Functionality Problem Report' generated."

Chapter 2

Getting Started

This chapter provides pointers to the various AIMMS examples that are available online and may help you to get a quick feel for the system. It explains the principle steps involved in creating a new AIMMS project, and it provides you with an overview of the various graphical tools available in AIMMS to help you create a complete Analytic Decision Support application. In addition, the chapter discusses the files related to an AIMMS project.

This chapter

2.1 Getting started with AIMMS

For most people, learning to use a new software tool like AIMMS is made substantially easier by first getting to see a few examples of its use. In this way you can get a quick feel for the AIMMS system, and begin to form a mental picture of its functionality.

Learn by example

In addition, by taking one or more illustrative examples as a starting point, you are able to quickly create simple but meaningful AIMMS projects on your own, without having to read a large amount of documentation. Building such small projects will further enhance your understanding of the AIMMS system, both in its use and in its capabilities.

Getting started quickly

To get you on your way as quickly as possible, the AIMMS system comes with a tutorial consisting of

AIMMS tutorial

- a number of live demos illustrating the basic use of AIMMS,
- an extensive library of small example projects each of which demonstrates a particular component of either the AIMMS language or the end-user interface,
- a number of complete AIMMS applications, and
- worked examples corresponding to chapters in the book on optimization modeling.

Example projects

The library of small example projects deals with common tasks such as

- creating a new project,
- building a model with your project,
- data entry,
- visualizing the results of your model,
- case management, and
- various tools, tips and tricks that help you to increase your productivity.

What you learn

By quickly browsing through these examples, you will get a good understanding of the paradigms underlying the AIMMS technology, and you will learn the basic steps that are necessary to create a simple, but fully functional modeling application.

This User's Guide

Rather than providing an introduction to the use of AIMMS, the User's Guide deals, in a linear and fairly detailed fashion, with all relevant aspects of the use of AIMMS and its modeling tools that are necessary to create a complete modeling application. This information enables you to use AIMMS to its full potential, and provides answers to questions that go beyond the scope of the example projects.

The Language Reference

The Language Reference deals with every aspect of AIMMS data types and the modeling language. You may need this information to complete the attribute forms while adding new identifiers to your model, or when you specify the bodies of procedures and functions in your model.

The Optimization Modeling guide

The Optimization Modeling guide provides you with the basic principles of optimization modeling, and also discusses several advanced optimization techniques which you may find useful when trying to accomplish nontrivial optimization tasks.

Context sensitive help

You can get online help for most of the tools, attribute forms and objects within the AIMMS system through the **Context Help** button 💡 on the AIMMS toolbar. It will point you to the appropriate section in one of the AIMMS books or in one of the help files that accompany the system.

How to proceed

The following strategy may help you to use AIMMS as efficiently and quickly as possible.

- Study some of the working examples to get a good feel for the AIMMS system.
- Select an example project that is close to what you wish to achieve, and take it as a starting point for your first modeling project.
- Use context help if you encounter a component of the AIMMS system that you are not familiar with, and are in need of a quick pointer providing some basic information.

■ Consult any of the three AIMMS books whenever you need more thorough information about either the use of AIMMS, its language or tips on optimization modeling.

2.2 Creating a new project

Every AIMMS project consists of two main components:

Project components

■ an AIMMS project file (with a .prj extension), which contains all project settings as well as all user-interface related components in the project such as end-user pages and menus, and

■ an AIMMS model file (with an .amb extension), containing all identifier declarations, procedures and functions that make up the core of your model.

When you start a new AIMMS project, you can either let AIMMS create both components (as explained below), or select an existing .prj or .amb file.

Within an AIMMS session, you can create a new project file through the **File-New Project** menu. Note that this menu is only available when no other project is currently open. It will open the AIMMS **New Project** wizard illustrated in Figure 2.1. In this wizard you can enter the name of the new project, along

Creating a new project file

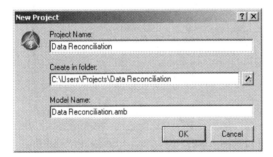

Figure 2.1: The AIMMS **New Project** wizard

with the directory in which the project is to be stored, and the model file (with the .amb extension) to be associated with the project.

By default, the AIMMS **New Project** wizard suggests that the new project be created in a new subdirectory with the same name as the project itself. You can use the wizard button ▨ to the right of the directory field to modify the location in which the new project is created. However, as AIMMS creates a number of additional files and directories in executing a project, you are strongly advised to store each AIMMS project in a separate directory.

Project directory

Model file

By default, the AIMMS **New Project** wizard assumes that you want to create a new model file with the same name as the project file (but with a different extension). You can modify the name suggested by the wizard to another existing or nonexisting model file. If the model associated with the project does not yet exist, it will be automatically created by AIMMS.

The Model Explorer

After you have finished with the **New Project** wizard, AIMMS will open the **Model Explorer**, an example of which is illustrated in Figure 2.2. The Model Ex-

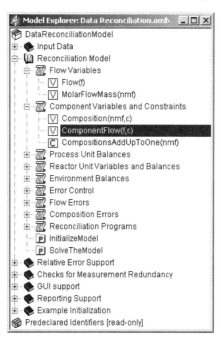

Figure 2.2: The AIMMS Model Explorer

plorer is the main tool in AIMMS to build an AIMMS model, the starting point of building any AIMMS application. In the Model Explorer, the model is presented as a tree of identifier declarations, allowing you to organize your model in a logical manner and make it easy—both for you and others who have to inspect your model—to find their way around. Besides the Model Explorer, AIMMS provides a number of other development tools for model building, GUI building and data management. An overview of these tools is given in Section 2.3.

Creating a project in the Windows Explorer

As an alternative to creating a new project file inside AIMMS itself, you can directly create a new AIMMS project from within the Windows Explorer. After you have installed AIMMS on your machine, the right-mouse pop-up menu in the right pane of the Windows Explorer window allows you to insert a new AIMMS project file into any directory via the **New-AIMMS Project File** menu. This will start a new AIMMS session, and directly open the **New Project** dialog box discussed above to create a new AIMMS project in the selected directory.

You can open an existing AIMMS project in two ways. You can either

- start AIMMS and open the project via the **File-Open Project** menu, or
- double click on the AIMMS project file (with a .prj extension) in Windows Explorer.

After opening a project, AIMMS may take further actions (such as automatically opening pages or executing procedures) according to the previously stored project settings.

Starting an existing project

2.3 Modeling tools

Once you have created a new project and associated a model file with it, AIMMS offers a number of graphical tree-based tools to help you further develop the model and its associated end-user interface. The available tools are:

Modeling tools

- the *Model Explorer*,
- the *Identifier Selector*,
- the *Page Manager*,
- the *Template Manager*,
- the *Menu Builder*,
- the *Data Manager*, and
- the *Data Management Setup* tool.

These tools can be accessed either through the **Tools** menu or via the project toolbar. They are all aimed at reducing the amount of work involved in developing, modifying and maintaining particular aspects of your model-based end-user application. Figure 2.3 provides an overview of the windows associated with each of these tools.

The AIMMS **Model Explorer** provides you with a simple graphical representation of all the identifiers, procedures and functions in your model. All relevant information is stored in the form of a tree, which can be subdivided into named sections to store pieces of similar information in a directory-like structure. The leaf nodes of the tree contain the actual declarations and the procedure and function bodies that make up the core of your modeling application. The **Model Explorer** is discussed in full detail in Chapter 4.

The Model Explorer

While the **Model Explorer** is a very convenient tool to organize all the information in your model, the **Identifier Selector** allows you to select and simultaneously view the attributes of groups of identifiers that share certain functional aspects in your model. By mutual comparison of the important attributes, such overviews may help you to further structure and edit the contents of your model, or to discover oversights in a formulation. The **Identifier Selector** is discussed in full detail in Chapter 7

The Identifier Selector

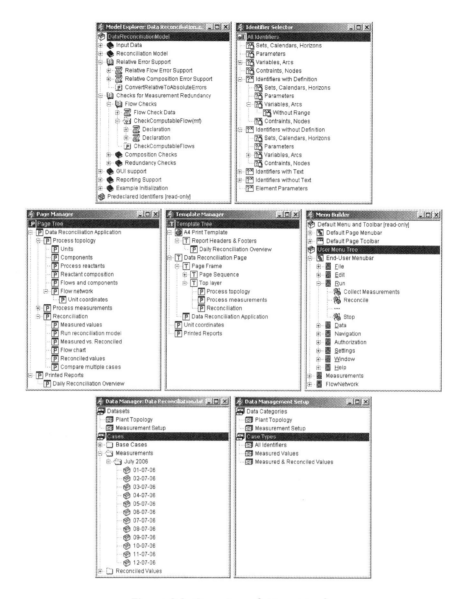

Figure 2.3: Overview of AIMMS tools

*The Page
Manager*
The **Page Manager** allows you to organize all end-user windows associated
with an AIMMS application (also referred to as end-user pages) in a tree-like
fashion. The organization of pages in the page tree directly defines the navigational structure of the end-user interface. Relative to a particular page in the
page tree, the positions of the other pages define common relationships such
as *parent* page, *child* page, *next* page or *previous* page, which can used in navigational controls such as buttons and menus. The **Page Manager** is discussed
in full detail in Section 12.1.

Within the **Template Manager**, you can make sure that all end-user pages have the same size and possess the same look and feel. You can accomplish this by creating page templates which define the page properties and objects common to a group of end-user pages, and by subsequently placing all end-user pages into the tree of page templates. The **Template Manager** is discussed in full detail in Section 12.2.

The Template Manager

With the **Menu Builder** you can create customized menu bars, pop-up menus and toolbars that can be linked to either template pages or end-user pages in your application. In the menu builder window you can define menus and toolbars in a tree-like structure similar to the other page-related tools, to indicate the hierarchical ordering of menus, submenus and menu items. The **Menu Builder** is discussed in full detail in Section 12.3.

The Menu Builder

AIMMS offers an advanced scheme for storing model results and dealing with multiple scenarios through the use of cases and datasets. With the **Data Manager** you can manage the entire collection of cases and datasets constructed for a particular AIMMS application. In addition, you can use the **Data Manager** to initiate a batch run for a set of cases or to view model data for a selection of cases. The **Data Manager** is discussed in full detail in Chapters 16 and 17.

The Data Manager

With the **Data Management Setup** tool you can specify the types of cases and datasets that are relevant for your application. Such a subdivision may help your end-users to store only the information necessary during a certain stage of your model, or to share common data between multiple cases. The **Data Management Setup** tool is discussed in full detail in Chapter 17.

The Data Management Setup tool

2.4 Dockable windows and tabbed MDI mode

As of AIMMS release 3.4, AIMMS supports the use of dockable windows for the tools discussed in the previous section. Dockable windows are an ideal means to keep frequently used tool windows in an development environment permanently visible, and are a common part of modern Integrated Development Environments (IDE) such as *Visual Studio .NET*.

Support for dockable windows

Dockable windows can be used in a *docked*, *auto-hidden*, or *floating* state. Whether a dockable window is in a docked, auto-hidden or floating state can be changed at runtime through their system context menu.

Docking states

When docked, the tool windows are attached to the left, right, top or bottom edge of the client area of the main AIMMS window. By default, all modeling tools discussed in Section 2.3 are docked at the left edge of the AIMMS window,

Docked windows

as illustrated in Figure 2.4. When you open a second instance of the same tool it is docked at the right edge of the AIMMS window.

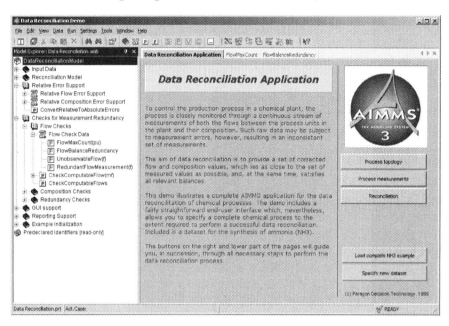

Figure 2.4: Example of dockable windows and tabbed MDI mode

Dragging docked windows

By dragging the windows caption of a docked window and moving the cursor around the edges of the AIMMS window, you can move the docked window to another position. A drag rectangle snaps into place at the appropriate edge, whenever a dockable window is ready to be docked at the location where you moved it. By moving the windows caption into the window area of another, dockable windows can also be docked to each other as tabs. Figure 2.4 illustrates the **Model Explorer** and **Page Manager** docked to each other as tabs. The area of a docked window can also be split into two by dragging another dockable window into the upper, lower, left or right part of docked window. In all these cases, a drag rectangle shows how a dockable window will be docked when you release the mouse at that time.

Auto-hidden windows

In auto-hidden state, a dockable window is normally collapsed as a button to the upper, lower, left or right edge of the main AIMMS window. By default, the **Message/Error** window of an AIMMS application, is collapsed to the bottom edge of the main AIMMS window. When you push the button associated with a collapsed window, it is expanded. When an expanded tool window looses focus, it is collapsed again. By clicking the pushpin button 🔲 in the caption of a docked/collapsed window, you can change the window's state from docked to auto-hide and back. Whenever a runtime error occurs, the **Message/Error** window will be activated and expand. If you do not close it, the window will collapse automatically as soon as you continue working in any other window. If another error occurs, the window will re-expand automatically.

By dragging a tool window away from an edge of the main AIMMS window, it becomes floating. When AIMMS is the active application, floating tool windows are always visible on top of all other (non-floating) AIMMS windows.

Floating tool windows

In addition to dockable windows, AIMMS 3.4 supports *tabbed MDI* mode for all other (document) windows such as attribute windows, identifier views and pages. In Figure 2.4 the main page of the application is displayed in tabbed MDI mode, with the attribute windows of two identifiers in the model accessible through tabs. Tabbed MDI windows occupy the remaining space of the client area of the main AIMMS window that is not occupied by docked windows. This implies that you do not have control over the size of tabbed MDI windows. Therefore, if you use tabbed MDI mode in your AIMMS application, it makes sense to make all the pages in your model resizable (see Chapter 13).

Tabbed MDI mode

When you drag the tab associated with a document window in the document window area, you can move the document window into a new or existing *tab group*, at the left, right, top or bottom of the current tab group. Tab groups effectively split the document window area into multiple parts, each of which can hold one ore more tabbed MDI windows. As with dragging dockable windows, a drag rectangle shows where the window will be positioned if you drop it at that moment. Tab groups are very convenient, for instance, if you want to view two attribute windows simultaneously.

Tab groups

By default, AIMMS uses dockable windows and tabbed MDI style document windows when it starts up. If you do not wish to use these features, you can change the style of the AIMMS IDE during an AIMMS session, through the **Windows- IDE Style** menu. If you uncheck the **Use Docking Windows** item, all tool windows that you open from then will will be opened as ordinary document windows. If you uncheck the **Use Tabbed MDI Mode** item, AIMMS will display all document windows as ordinary MDI document windows. Thus, if you uncheck both features, all AIMMS windows will be displayed as ordinary MDI document windows, compatible with the window display mode of AIMMS 3.3 and earlier.

Change IDE style

You can also change the IDE style permanently, on a per-project basis. Through the project options contained in the **Options-Project-IDE Style** area (see also Section 21.1), you can permanently change the IDE style settings for a specific project. If a project is not suitable for tabbed MDI mode by design (e.g. because it contains a lot of small pages that cannot be displayed properly in full screen size tabbed MDI windows), you can also disable the **Windows-IDE Style** menu item for such a project. Thus, you can prevent your end-users from changing

IDE style project options

the IDE style of Aimms if that is undesirable for your project.

2.5 Files related to an Aimms project

Project-related files

A number of files are associated with every Aimms project. Some of these files are necessary to run your project, while others are optional or may be generated automatically by Aimms. The following file types can occur:

- the main project file (with a .prj extension),
- one or more model source files (with a .amb extension),
- the name change file (with a .nch extension),
- one or more data manager files (with a .dat extension),
- a user database file (with a .usr extension),
- one or more library project files (with a .libprj extension),
- encrypted model source files (with a .aeb extension),
- project and data backup files (with a .bak extension),
- ASCII backup files of the model source (with a .aim extension), and
- log, error and listing files from both Aimms and its solvers (with .log, .err, .lis or .sta extensions).

The main project file

The main project file (with the .prj extension) contains the general setup information related to the project, such as the name of the model source file or user database associated with the project, the project options and settings, as well as the data management setup. In addition, the project file contains the specification of all the end-user pages, templates and menus.

The model source

A model source file (with a .amb extension) contains the source code associated with a project in a binary format. The source code of a project may be separated into several .amb files, which can be included in the main model file. This allows multiple developers to work on the source of a single project, or to use common declarations and procedures within multiple projects.

Name change file

Aimms has the capability to keep track of the name changes you performed on identifiers in your model, and automatically replace an old identifier name by its new one whenever Aimms encounters a renamed identifier. Aimms keeps track of the list of such name changes in the name change file (with a .nch extension). Each name change is listed in this file on a separate line containing the old name, the new name and the (GMT) time stamp at which the change was actually performed. The automatic name change capabilities of Aimms are explained in full detail in Section 5.3

Data manager files (with a .dat extension) contain a set of cases and datasets created in your AIMMS model-based application. By default, AIMMS will create and open a single data manager file with the same name as the project file. You can open other data manager files (containing other cases and datasets) through the **File-Open** menu.

Data manager files

With every end-user project created with AIMMS, you can associate an end-user database containing the names and passwords of all end-users entitled to run the project or view its results. Such end-user information is stored in an encrypted format in a user database file (with a .usr extension). You can use a single user database file with more than one project.

The user database

One or more library projects can be added to your project. A library project can be seen as a complete project (including model source, templates, end-user pages and menus) that can be accessed from within the main project. The .libprj file is, as its name might imply, the project file of a library. Libraries in AIMMS make it easier to work with multiple developers on a single project. A detailed description of all multi-developer features can be found in Chapter 3.

Libary project files

When shipping your project to an end-user you can choose protect your model by encrypting the model source into a *one way encrypted* model. Such an encrypted model file will have the extension .aeb and cannot be viewed by the end-user. The process of generating a one way encrypted model is described in detail in the chapter on project security (see Chapter 20, Section 20.1).

Encrypted source files

During the execution of your model, all log, error and listing information from both AIMMS and its solvers (whether visible in the AIMMS **Message** window or not) is copied to log, error and listing files, which, by default, are stored in the Log subdirectory of the project directory. If you are not interested in this information, you can reduce the amount of information that is copied to these log files by modifying the relevant execution options.

Log, error and listing files

2.5.1 Project backup files

When you are developing an AIMMS model, AIMMS will create regular back-ups (with a .bak extension) of the project files and data associated with your project. By default, such backup files will be stored in the Backup subdirectory of the project directory. In addition to storing a backup file of the binary source file (with the .amb extension), most AIMMS versions can also create ASCII backup files of the model source (with a .aim extension) for your convenience.

Backup files

Project backups   You can modify AIMMS' default project backup settings through the **AutoSave & Backups-Project** menu, which will pop up the **Project AutoSave & Backups** dialog box illustrated in Figure 2.5. In this dialog box, you can specify

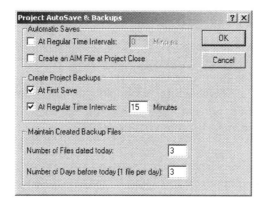

Figure 2.5: The **Project AutoSave & Backups** dialog box

- whether you want AIMMS to automatically save your project at regular time intervals,
- whether you want AIMMS to create an ASCII .aim file of your model when you close the project,
- at which times you want AIMMS to store a backup of your project, and
- how many project backups AIMMS will retain.

Creating .aim files   When you have checked the .aim file creation check box, AIMMS will create a .aim file of your model whenever you close the project. You can manually create .aim files of your model through the **File-Save As** menu in the **Model Explorer**. The .aim file contains an equivalent ASCII representation of the contents of the model tree, mainly intended for your convenience.

Loading .aim files   If you so desire (e.g. after making manual edits to the .aim file), you can restore the model tree from a .aim file through the **File-Open- Model** menu. When you edit a model loaded from a .aim file, AIMMS will automatically convert it back to a binary .amb file when you save the model for the first time.

Backup policy   In the **AutoSave & Backups** dialog box, you can indicate at which times you want AIMMS to create backups of your project. By default, AIMMS will save such backup files (with the .bak extension) in the Backup directory of the project directory. You can modify these settings in the **Project-Directories** folder of the **AIMMS Options** dialog box (see Section 21.1). A backup file contains a copy of both the project and the model file, and its name contains a reference to the date/time of the backup. To keep the number of backup files down to a reasonable number, the dialog box of Figure 2.5 allows you to indicate how many backup files AIMMS will retain when closing the project.

Through the **AutoSave & Backups-Data** menu, you can specify that you want   *Data backups*
AIMMS to automatically create backups of the data used during a particular
session of your project. The menu will pop up the **Data Backup** dialog box
illustrated in Figure 2.6. Similarly as with the project backup files, you can

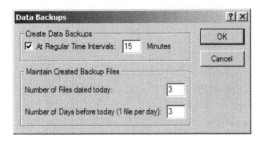

Figure 2.6: The **Data Backup** dialog box

indicate whether AIMMS should automatically create backup backup files of
the session data at regular intervals, as well as how many data backup files
should be retained. Data backup files also have the .bak extension and contain
a reference to the date/time of the backup.

Besides the automated backup scheme built into AIMMS, you can also create   *Manually*
backup files of either your project or the session data manually. You can   *creating backup*
create manual backup files through the File-Backups-Project-Backup and File-   *files*
Backups-Data-Backup menus. When you create a project or data backup file
manually, AIMMS will request a name of a .bak file in which the backup is to be
stored.

Through the File-Backups menu, you can also restore project and data backup   *Restoring*
files. When you decide to restore the project files, AIMMS will request you to   *backup files*
select a project backup file to restore from, after which you get an overview
of the file differences (compared to the current project files) in the **Restore**

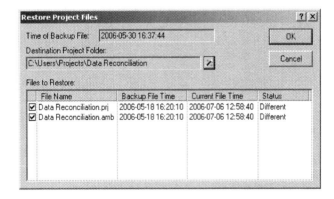

Figure 2.7: The **Restore Project Files** dialog box

Project Files dialog box illustrated in Figure 2.7. In this dialog box, you can indicate which parts of the project backup you want to be restored. This allows you, for instance, to selectively restore the model file, but retain the current project file.

2.5.2   Project user files

Project user files

Along with the project-related files created by AIMMS, you may need to distribute some other files with your project. Such files include, for instance, bitmap files displayed on buttons or in the background of your end-user pages, or files that contain project-related configuration data. Instead of having to include such files as separate files in the project directory, AIMMS also allows you to save them within the project file itself. Both within the AIMMS language as well as in the end-user interface, you can reference such *project user files* as if they were ordinary files on disk.

Why use project user files?

User project files are convenient in a number of situations. The most common reasons to store files as project user files are listed below.

- You want to reduce the number files that you have to ship to your end users. This situation commonly occurs, for instance, when the end-user interface of your project references a large number of bitmap files.
- You want to hide particular configuration data files from your end-users, which might otherwise only confuse them.

Importing project user files

You can import files into the project file through the **Tools-Project User Files** menu, which will pop up the **Project User Files** dialog box illustrated in Figure 2.8. In this dialog box, you can create new folders to organize the files

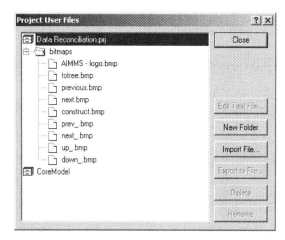

Figure 2.8: The **Project User Files** dialog box

you want to import into the project file. The dialog box of Figure 2.8 already contains a folder **bitmaps**, which is automatically added to each new AIMMS project and filled by AIMMS with the bitmaps used on AIMMS' data pages (see Section 5.4). When you are inside a folder (or just within the main project file), you can import a file into it through the **Import File** button, which will open an ordinary file selection dialog box to select the disk file to be imported.

When your project, next to the main project file, also includes a number of library project files (see Section 3.1), AIMMS allows you to store user files in the library project files as well. Thus, if a page defined in a library refers to a particular bitmap file, you can also store that bitmap as a user file directly into the corresponding library project file. In the dialog box of Figure 2.8, the *CoreModel* node at the root of the tree refers to a library that is included in the project that serves as the running example throughout this book. Underneath this node you can add user files that will be stored in the library project file for the *CoreModel* library.

User files in library projects

When you import a bitmap file (with the .bmp extension) into the project file, AIMMS will compress its contents before storing it into the project file (and decompress it before referencing its contents). Such compression may lead to a considerable storage reduction when you are importing large bitmap files, and forms an additional argument to store bitmap files used within the end-user interface into the project file.

Bitmap files are compressed

You can reference project user files both from within the AIMMS language and the properties of various objects with the graphical end-user interface. The basic rule is that AIMMS considers the project file as a virtual disk indicated by "<prj>". You can use this virtual drive in, for instance, READ, WRITE and PUT statements within your model. Thus, the statement

Referencing project user files

```
READ from file "<prj>:config\\english.dat";
```

reads the model data from the project user file "english.dat" contained in a (developer-created) **config** folder within the project file.

You can access project files in library projects by using the virtual disk notation "<lib:*library-name*>", where *library-name* is the name of the library project. Thus, to read the same file as in the previous paragraph from the *CoreModel* library shown in Figure 2.8, the following statement can be used.

Referencing user files in library projects

```
READ from file "<lib:CoreModel>:config\\english.dat";
```

Use in end-user
interface

Similarly, you can reference project user files on page objects in the end-user interface of your project. Figure 2.9 illustrates the use of a bitmap file stored in the project file on a bitmap button. For all object properties expecting a file

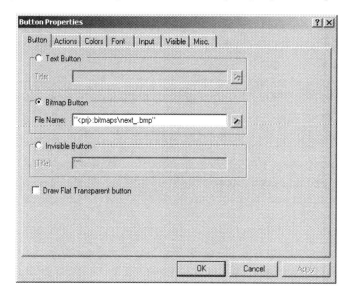

Figure 2.9: Bitmap button referencing a project user file

name (such as the **File Name** property of the bitmap button illustrated in Figure 2.9), you can easily select a project user file by pressing the wizard button and selecting the **Select Project File** menu item. This will pop up a project user file selection dialog box similar to the dialog box shown in Figure 2.8.

2.6   AIMMS 3 licensing

AIMMS 3
licensing

AIMMS offers the following two types of licenses:

- single-user licenses, and
- network licenses.

Each of these two types of licenses are protected in a different manner.

Single-user
license
protection

Single-user licenses can be used by a single user on a single computer. To enforce the single-user character, AIMMS requires that single-user licenses be protected by either

- a hardware dongle, which, depending on the type of dongle, must be connected to a USB or parallel port of your computer, or
- a nodelock file, which must be activated to match the hardware characteristic of your computer.

When ordering the Win32 version of the AIMMS software you can indicate whether you want your AIMMS system to be protected by a dongle or by a nodelock. Which choice to make is very dependent on your situation and the intended use of the AIMMS software. The Win64-IA64 and Win64-x64 platforms do not support dongles.

You can choose

Dongles offer you the most flexibility when you want to use AIMMS on multiple computers, and do not want the hassle of having to deactivate and activate a nodelock on these computers. On the other hand, dongles occasionally break, you can forget to take the dongle with you, they can be stolen, and, because of their size, get lost quite easily, especially if you are moving them around a lot.

Pros and cons of dongles

Nodelock files are stored on the harddisk of your computer, and are, therefore, much less vulnerable to loss. Only if you computer is stolen, or in case of a harddisk crash, you must contact Paragon before being able to activate your nodelock on a replacement computer. On the other hand, if you are frequently working on multiple computers, you have to remember to deactivate the nodelock on the old computer, prior to being able to activate it on the new one, every time. In addition, you need access to the internet to activate or deactivate a nodelock.

Pros and cons of nodelocks

If you decide to request a dongle for license protection, a physical shipment of the dongle to your site is required before you can start using AIMMS. If you request nodelock protection, we will send you the AIMMS license number and activation code by e-mail, after which you can start using AIMMS directly.

Physical shipments

If you request your license to be protected by a dongle, an AIMMS dongle is sent to you along with your AIMMS 3 CD-ROM. If you upgrade from AIMMS 2.20, you can continue to use your existing AIMMS 2.20 hardware dongle. The green Activator as well as the grey Sentinel dongle must be connected to the parallel port of your computer. The purple Sentinel dongle must be connected to a USB port of your computer. The AIMMS setup program will only install the required device drivers for accessing the grey and purple Sentinel dongles. If you still use the green Activator dongles supplied with AIMMS 2.20, you can obtain the required drivers separately from our FTP site ftp.aimms.com.

AIMMS dongles

If you have ordered an AIMMS 3 network license, no license protection needs to be installed locally on your computer. Instead, you need the host name and port number of the server running the AIMMS 3 network license server. For more information about installing the network license server itself, please refer to the documentation of the AIMMS 3 network license server.

Network licenses

2.6.1   Personal and machine nodelocks

Two types of nodelocks

AIMMS offers two types of nodelocks:

- personal nodelocks, and
- machine nodelocks.

If you choose for nodelock protection you are free to choose between a personal or a machine type of nodelock. In this section you will find the characteristics of both types of nodelocks. If you are unsure which type of nodelock to choose, we recommend that you start with a personal nodelock, as you can change a personal nodelock into a machine nodelock at any time, but not the other way around.

Personal nodelock

Personal nodelocks are intended for use by a single AIMMS user, who still wishes to have the freedom to use AIMMS on multiple computers, for instance if you want to easily switch between your desktop computer at work, a notebook computer and your home computer. Personal nodelocks have the following characteristics:

- Personal nodelocks can be transferred to another computer 3 times per 24 hours. This allows you to take your AIMMS license home in the evening and back to work the next morning without any problems.
- Personal nodelocks have a limited lifetime of 60 days, and should be renewed within that period to extend the lifetime to its full 60-day period. If the nodelock is not renewed within its 60-day lifetime, this does not invalidate your AIMMS license in any way—you only have to renew your nodelock prior to being able to use your AIMMS system again. Note that the renewal process does not require any manual intervention, as AIMMS will try to automatically connect to our internet license database to renew your nodelock once every day you are using AIMMS.
- Both activation and nodelock renewal of personal nodelocks require a working connection to the internet. As a consequence, in the absence of an internet connection you can continue to work uninterrupted for a period of 60 days, before an internet connection is required to renew your nodelock.
- With every activation or nodelock renewal AIMMS will also update your license files if new license files are available (e.g. if your system is in maintenance), and will inform you of any messages that are available for you in our database.
- Because of their volatile nature, PDT will replace a personal nodelock without any questions asked in case of loss of or damage to your computer.
- You can switch your personal nodelock to a machine nodelock at any time.

Machine nodelocks are intended for permanent use on a single computer. They are recommended for server applications, and can also be used for personal use if you are sure you will be using AIMMS on a single computer, or do not have internet access. Machine nodelocks have the following characteristics:

Machine nodelock

- Machine nodelocks can be transferred to a replacement computer 3 times per 365 days.
- Machine nodelocks have an unlimited lifetime (unless deactivated).
- Machine nodelocks can be either activated online if your computer is connected to the internet, or offline through the license activation area on the AIMMS website.
- License files will only be retrieved when the machine nodelock is activated, or by explicit request.
- In case of failure, PDT will, in principle, only replace machine nodelocks on the same computer.
- Once you have chosen for a machine nodelock, it is not possible to switch back to a personal nodelock.

Although a personal nodelock makes a regular connection to our internet license database for nodelock renewal, we do respect your privacy and will not register patterns in your usage of the AIMMS software in any way. During activation no personal information will be transferred, only your computer name and some of its hardware characteristics. During deactivation we register the date and time of deactivation to enforce the transfer limit.

Privacy

The connection to our internet license database is implemented as a web service. Thus, if you are able to browse the web, you should also have no trouble activating an AIMMS nodelock. If your computer connects to the internet through a proxy server, AIMMS by default tries to detect and use the proxy settings also used by Microsoft Internet Explorer.

Internet connection and proxy settings

It should be noted that the use of auto-configuration scripts in determining the proxy server will fail if these use any other scripting language than Javascript. This is due to the libraries underlying the SOAP library used by AIMMS to connect to our license server. If you are in this situation, you should manually configure the proxy settings, as described below.

Automatic configuration scripts

If AIMMS does not succeed in automatically detecting the proxy settings that apply in your network environment, AIMMS also allows you to manually set the proxy settings during the activation process. If the online activation process does not succeed directly, AIMMS gives you the option to either continue with an offline activation process, or to manually supply the proxy settings that apply to your network environment through the dialog box illustrated in Figure 2.10. In this dialog box you can choose between

Manual proxy setting

- the *Current User* settings also used by Microsoft Internet Explorer (default),
- the Local Machine settings which are stored in the registry, if these are available on your machine, or
- *Custom* proxy settings that you have received from your IT department.

In the latter case, you can also (optionally) provide a user name and password to authenticate with the proxy server. In most cases, however, setting these will not be necessary, and Windows authentication will be sufficient.

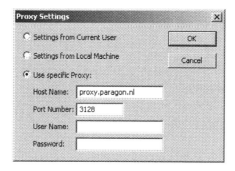

Figure 2.10: The AIMMS **Proxy Configuration** dialog box

2.6.2 Installing an AIMMS license

Managing your AIMMS licenses

When you start up AIMMS for the first time after installation, AIMMS will open the **License Configuration** dialog box illustrated in Figure 2.11. Through this dialog box you can install new AIMMS licenses and manage all AIMMS licenses that already have been installed on your computer.

Installing a new license

To install a new license, press the **Install License ...** button in the **License Configuration** dialog box. This will start a wizard, that will guide you through the license installation procedure step by step. The wizard can help you to install

- existing AIMMS licenses,
- nodelocked licenses,
- dongled licenses,
- network licenses,
- evaluation licenses, and
- student licenses.

To successfully complete the installation of licenses of each type, you should make sure to have the following information available.

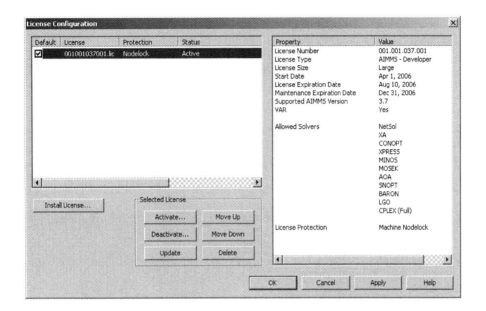

Figure 2.11: The **License Configuration** dialog box

To install a single-user AIMMS license that is protected by a nodelock, you need the following information:

- your AIMMS license number, and
- the associated activation code that you received from Paragon.

Single-user nodelocked licenses

You have the choice to request a personal nodelock or a machine nodelock. A personal nodelock must be requested online, a machine nodelock can be requested online or offline. Refer to Section 2.6.1 for a more detailed introduction to personal and machine nodelocks.

To install a single-user AIMMS license that is protected by a dongle, you need the following items:

- an AIMMS dongle attached to a USB or parallel port of your PC, and
- the associated set of license files that you received from Paragon.

Single-user dongled licenses

To install an AIMMS network license, you need the following information from your system administrator:

- the name of the AIMMS network license server,
- the port number of the AIMMS network license server, and
- the name of the license profile to which you want to connect (optional).

Network licenses

*Evaluation
licenses*

To install an AIMMS evaluation license you need the following information

- your AIMMS evaluation license number, and
- the associated activation code that you received from Paragon when requesting an evaluation license.

You must have a working connection to the internet to activate an evaluation license. Evaluation licenses expire 30 days after activation. Note that each evaluation license can be activated only once, and that you can only activate a single evaluation license per AIMMS release on a specific computer, regardless of the number of evaluation licenses you have requested on our web site.

Student licenses

To install an AIMMS student license you need the following information:

- your AIMMS student license number, and
- the associated activation code that you received from the university that purchased the AIMMS Educational Package.

You must have a working connection to the internet to activate a student license. Student licenses expire one month after the end of the current academic year. Student licenses can be activated multiple times.

2.6.3 Managing AIMMS licenses

*Managing
multiple* AIMMS
licenses

AIMMS allows you to have multiple AIMMS licenses installed on your computer. You may have multiple licenses installed, for instance, for the following reasons:

- you have requested a trial license for a new AIMMS version which you want to run next to your existing license,
- you have temporarily borrowed or hired an AIMMS license with more capabilities than your regular license,
- your system administrator has created multiple network license profiles, each of which you may want to use to run AIMMS.

In this section we will describe how you can instruct AIMMS which license to use.

Default licenses

In the **License Configuration** dialog box displayed in Figure 2.11, all AIMMS licenses installed on your machine will be displayed in the left pane of the dialog box. The license details of each license are displayed in the right pane of the dialog box. During startup AIMMS will consider all licenses in the left pane of the **License Configuration** dialog box which have the **Default** column checked, and will use the first valid license it finds starting from top to bottom. Using the **Move Up** and **Move Down** buttons you can change the order in which AIMMS will search the list.

Both personal and machine nodelocks can be transferred to other computers. Personal nodelocks can be transferred upto three times a day, allowing you to take your license with you wherever you want. Machine nodelocks can be transferred three times per year, to a computer replacing the computer on which the nodelock is currently installed. To transfer a nodelocked license, you must

Transferring licenses

- deactivate the nodelock on the currently active computer, and
- activate it on the computer to which you want to transfer the license.

You can deactivate an active nodelock using the **Deactivate** button in the **License Configuration** dialog box. Deactivation will only succeed if there is no conflict with the transfer limit for the given nodelock type. This makes sure that there will never be a problem activating a deactivated license. After successful deactivation the license will not be removed from the list but be marked as inactive. If the license is not active on any computer, you can reactivate the license through the **Activate** button.

In case you want to activate a nodelock on a computer, but have forgotten to deactivate the nodelock on a computer to which you currently have no access, AIMMS allows you, as a courtesy, to request an emergency nodelock 3 times per 365 days. Emergency nodelocks have a lifetime of 3 days, and during this time you can arrange for someone to deactivate the license on the computer containing the active nodelock. During the activation sequence, AIMMS will automatically ask whether you would like to receive an emergency nodelock when it discovers that the license is active on another computer.

Emergency nodelocks

2.6.4 Location of license files

AIMMS keeps its license and configuration files in the folder

Location of license files

    Paragon Decision Technology

of the common application area of your computer. On Windows 2000, Windows XP, and Windows Server 2003, the common application area is located, by default, at

    C:/Documents and Settings/All Users/Application Data.

On Windows Vista, this folder is located under C:/ProgramData. The Paragon Decision Technology folder contains three subfolders

- Config, containing the license and solver configuration files,
- Licenses, containing all license files, and
- Nodelocks, containing all nodelock files installed on your computer.

The AIMMS installation makes sure that these subfolders are writable for everyone, allowing you to install and uninstall licenses on your computer.

Do not move nodelock files

To prevent tampering with nodelocked licenses, AIMMS keeps track of the location of the nodelock files associated with a license. You should, therefore, not manually move or copy the AIMMS nodelock files as this may invalidate your nodelock.

Personalization

During the installation of a license, AIMMS will install the license and solver configuration files that came with the license in the common application area, available to anyone who wants to use AIMMS on your computer. When you make changes to the license or solver configuration later on, however, such modifications will only be stored in your user application area, overriding the default configuration. In this manner, every user can create his own AIMMS configuration without bothering other users.

Chapter 3

Organizing a Project into Libraries

This chapter discusses the options you have in AIMMS to organize a project in such a way that it becomes possible to effectively work on the project with multiple developers. While it is very natural to start working on a project with a single developer, at some time during the development of an AIMMS application, the operational requirements of the problem you are modeling may become so demanding that it requires multiple developers to accomplish the task.

This chapter

During the initial prototyping phase of a model, an AIMMS project is usually still quite small, allowing a single developer to take care of the complete development of the prototype. The productivity tools of AIMMS allow you to quickly implement different formulations and analyze their results, while you are avoiding the overhead of having to synchronize the efforts of multiple people working on the prototype.

From proto-typing phase...

During the development of an operational AIMMS application this situation may change drastically. When an AIMMS application is intended to be used on a daily basis, it usually involves one or more of the following tasks:

...to operational phase

- retrieving the input data from one or more data sources,
- validation and transformation of input data,
- extending the core optimization application with various, computationally possibly demanding, operational requirements,
- preparing and writing output data to one or more data sources,
- building a professionally looking end-user GUI, and/or
- integrating the application into the existing business environment.

Depending on the size of the application, implementing all of these tasks may become too demanding for a single developer.

The most sensible approach is then to divide the project into several logical sub-projects, either based on the tasks listed in the previous paragraph, or more closely related to the logic of your application. All of these tasks are characterized by being rather self-contained, while, using the productivity tools of AIMMS, they can still be easily managed by a single developer.

Dividing a project into sub-projects

3.1 Library projects and the library manager

AIMMS library projects

AIMMS *library projects* allow you to divide a large AIMMS project into a number of smaller sub-projects. Each library project in AIMMS provides

- a tree of model declarations,
- a page tree,
- a template tree,
- a menu tree, and
- a data management setup tree.

In addition, a library project may provide its own collection of user project files, user colors and fonts.

Shared templates

Besides enabling multiple developers to work in a single project, library projects can also be used to define a common collection of templates that define the look-and-feel of multiple projects. In this manner, you change the look-and-feel of multiple applications just by changing the templates in the shared library project.

Adding libraries to a project

By adding a library project to the main AIMMS project, the objects defined by the library, such as identifiers, pages, templates, etc., become available for use in the main project. In addition, if a library project is writable, the contents of the library can also be edited through an AIMMS project in which it is included.

The library manager

You can add libraries to an AIMMS project in the **AIMMS Library Manager** dialog box illustrated in Figure 3.1. You can open the library manager through the **File-Library Manager...** menu.

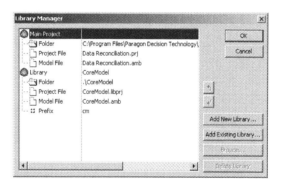

Figure 3.1: The **AIMMS Library Manager**

Using the library manager AIMMS allows you to

- create new libraries,
- add existing libraries to your project,

- edit library properties, and
- remove libraries from your project.

Each library project in AIMMS consists of two files: *Library files*

- a library project file (with the .libprj extension), and
- a library model file (with the .amb extension).

These files will be automatically created by AIMMS when you create a new library project. To add an existing library to an AIMMS project, you just need to select its library project file.

To avoid name clashes between objects in the library and the main project, all   *Library prefix*
the object names in a library are stored in a separate namespace. Outside of
the library, a global prefix associated with the library has to be used to access
the library objects. When you create a new library project, AIMMS will come
up with a default library prefix based on the library name you entered. For an
existing library project, you can view and edit its associated library prefix in
the library manager.

After you have added one or more library projects to your main AIMMS project,   *Using library*
AIMMS will extend   *projects*

- the model tree in the **Model Explorer**,
- the page tree in the **Page Manager**,
- the template tree in the **Template Manager**,
- the menu tree in the **Menu Builder**, and
- the data management setup tree in the **Data Management Setup** tool

with additional root nodes for every library project added to your project.
In general, within any of these tools, you are free to move information from
the main project tree to any of the library trees and vice versa. In addition,
the AIMMS dialog boxes for user project files, user colors and fonts allow you
to select and manage objects from the main project or any of the libraries.
The precise details for working with library projects in each of these tools are
discussed in full detail in the respective chapters discussing each of the tools.

3.2 Guidelines for working with multiple developers

Unless you started using library projects from scratch, you need to convert   *Identifying*
the contents of your AIMMS project as soon as you decide to divide the project   *independent*
into multiple library projects. The first step in this process is to decide which   *tasks*
logical tasks in your application you can discern that meet the following criteria:

- the task represents a logical unit within your application that is rather self-contained, and can be comfortably and independently worked on by a single developer at a time, and
- the task provides a limited interface to the main application and/or the other tasks you have identified.

Good examples of generic tasks that meet these criteria are the tasks listed on page 37. Once your team of developers agrees on the specific tasks that are relevant for your application, you can set up a library project for each of them.

Library interface...

The idea behind library projects is to be able to minimize the interaction between the library, the main project and other library projects. At the language level Aimms supports this by letting you define an *interface* to the library, i.e. the set of public identifiers and procedures through which the outside world is allowed to connect to the library. Library identifiers not in the interface are strictly private and cannot be referenced outside of the library. The precise semantics of the interface of a library module is discussed in Section 31.5 of the Language Reference.

... used in model and GUI

This notion of public and private identifiers of a library module does not only apply to the model source itself, but also propagates to the end-user interface. Pages defined in a library can access the library's private identifiers, while paged defined outside of the library only have access to identifiers in the interface of the library.

Minimal dependency

The concept of an interface allows you to work independently on a library. As long as you do not change the declaration of the identifiers and procedures in its interface, you have complete freedom to change their implementation without disturbing any other project that uses identifiers from the library interface. Similarly, as long as a page or a tree of pages defined in a library project is internally consistent, any other project can add a reference to such pages in its own page tree. Pages outside of the library can only refer to identifiers in the library interface, and hence are not influenced by changes you make to the library's internal implementation.

Conversion to library projects

If your application already contains model source and pages associated with the tasks you have identified in the previous step, the next step is to move the relevant parts of your Aimms project to the appropriate libraries. You can accomplish this by simply dragging the relevant nodes or subtrees from any of the trees tree in the main project to associate tree in a library project. What should remain in the global project are the those parts of the application that define the overall behavior of your application and that glue together the functionality provided by the separate library projects.

3.3 Version control

Once you have accomplished the previous two steps, you can effectively start working on your AIMMS application with multiple developers. It is common in such situations to use some form of version control. If your version control software employs a lock-modify-unlock semantics, only files locked by you will be writable on disk, while all others are kept read-only. AIMMS honors these settings, by only allowing you to edit those parts of the project for which the associated disk files are writable.

Version control

Because AIMMS files are stored in binary format, employing a lock-modify-unlock semantics with your version control software is the most appropriate mode for versioning the AIMMS project and model files. These semantics are supported by both *Visual SourceSafe* (version control software by Microsoft) and *Subversion* (an open-source version control system).

Binary files

To ease tracking changes in your model, AIMMS enables you to also save the model in the ASCII .aim format, and even to automatically generate .aim files when you close the project (see Section 2.5.1). You can, therefore, decide to keep both the binary .amb and the ASCII .aim files under version control, and use the .aim files to keep track of the changes in your model.

Tracking model changes

AIMMS does not support an ASCII format for the project file, as it contains information for which no human-readable format is available. However, for pages in your project you can still view their last modification time. This enables you, in a limited sense, to track the changes in the end-user GUI of your AIMMS project. Through the **File-Open-Page** menu you can open the **Page Open** dialog illustrated in Figure 3.2.

Viewing page modification time

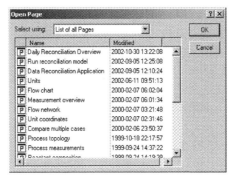

Figure 3.2: Viewing the modification time of pages

By selecting pages using the **List of all pages**, AIMMS will display the last modification time for every page in the project.

Part II

Creating and Managing a Model

Chapter 4

The Model Explorer

This chapter introduces the interactive **Model Explorer** that is part of the AIMMS system. With the **Model Explorer** you have easy access to every component of the source of your model. In this chapter, you are introduced to the model tree, and you are shown which model information can be added to the model tree. In addition, the basic principles of working with the **Model Explorer** are explained.

This chapter

4.1 What is the Model Explorer?

Decision making commonly requires access to massive amounts of information on which to base the decision making process. As a result, professional decision support systems are usually very complex programs with hundreds of (indexed) identifiers to store all the data that are relevant to the decision making process. In such systems, finding your way through the source code is therefore a cumbersome task. To support you in this process, AIMMS makes all model declarations and procedures available in a special tool called the **Model Explorer**.

Support for large models

The AIMMS **Model Explorer** provides you with a simple graphical model representation. All relevant information is stored in the form of a *model tree*, an example of which is shown in Figure 4.1. As you can see in this example, AIMMS does not prescribe a fixed declaration order, but leaves it up to you to structure all the information in the model in any way that you find useful.

Structured model representation

As illustrated in Figure 4.1, the model tree lets you store information of different types, such as identifier declarations, procedures, functions, and model sections. Each piece of information is stored as a separate node in the model tree, where each node has its own type-dependent icon. In this section, the main node types in the model tree will be briefly introduced. In subsequent chapters, the details of all model-related node types such as identifiers, procedures and functions will be discussed in further detail.

Different node types

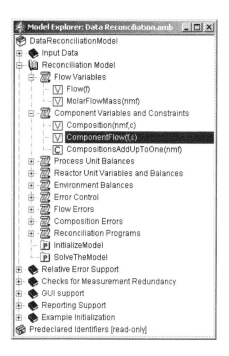

Figure 4.1: Example of a model tree

Structuring nodes

There are three basic node types available for structuring the model tree. You can branch further from these nodes to provide more depth to the model tree. These basic types are:

- The *main model* node which forms the root of the model tree. The main model is represented by a box icon 📦 which opens when the model tree is expanded, and can contain book sections, declaration sections, procedures and functions.
- *Book section* nodes are used to subdivide a model into logical parts with clear and descriptive names. Book sections are represented by a book icon 📖 which opens when the section is expanded. A book section can contain other book sections, declaration sections, procedures and functions.
- *Declaration section* nodes are used to group identifier declarations of your model. Declaration sections are represented by a scroll icon 📜, and can only contain identifier declaration nodes.

Advantages

The structuring nodes allow you to subdivide the information in your model into a logical framework of sections with clear and descriptive names. This is one of the major advantages of the AIMMS model tree over a straightforward ASCII model representation, as imposing such a logical subdivision makes it much easier to locate the relevant information when needed later on. This helps to reduce the maintenance cost of AIMMS applications drastically.

In addition to the basic structuring nodes discussed above, AIMMS supports two additional structuring node types, which are aimed at re-use of parts of a model and working on a single AIMMS project with multiple developers.

Module and library nodes

- The *module* node offers the same functionality as a book section, but stores the identifiers it defines in a separate namespace. This allows a module to be included in multiple models without the risk of name clashes. Module nodes are represented by the icon ⬢.
- The *library module* node is the source module associated with a library project (see Section 3.1). Library modules can only be added to or deleted from a model through the **Library Manager**, and are always displayed as a separate root in the model tree. Library module nodes are represented by the icon ⬢.

Modules, library modules and the difference between them are discussed in full detail in Chapter 31 of the Language Reference.

For your convenience, AIMMS always includes a single, read-only library module called Predeclared Identifiers (displayed in Figure 4.1), containing all the identifiers that are predeclared by AIMMS, categorized by function.

AIMMS library

All remaining nodes in the tree refer to actual declarations of identifiers, procedures and functions. These nodes form the actual contents of your modeling application, as they represent the set, parameter and variable declarations that are necessary to represent your application, together with the actions that you want to perform on these identifiers.

Non-structuring nodes

The most frequent type of node in the model tree is the identifier declaration node. All identifiers in your model are visible in the model explorer as leaf nodes in the declaration sections. Identifier declarations are not allowed outside of declaration sections. AIMMS supports several identifier types which are all represented by a different icon. The most common identifier types (i.e. sets, parameters, variables and constraints) can be added to the model tree by pressing one of the buttons ⬚⬚⬚⬚⬚ (the last button opens a selection list of all available identifier types). Identifier declarations are explained in full detail in Chapter 5.

Identifier nodes

Identifiers can be used independently of the order in which they have been declared in the model tree. As a matter of fact, you may use an identifier in an expression near the beginning of the tree, while its declaration is placed further down the tree. This order independence makes it possible to store identifiers where you think they should be stored logically, which adds to the overall maintainability of your model. This is different from most other systems where the order of identifiers is dictated by the order in which they are used inside the model description.

Independent order

Procedure and function nodes

Another frequently occurring node type is the declaration of a procedure or a function. Such a procedure or function node contains the data retrieval statements, computations, and algorithms that make up the procedural execution of your modeling application. Procedures and functions are represented by folder icons, 🄿 and 🄵, which open when the procedure or function node is expanded. They can be inserted in the model tree in the root node or in any book section. The fine details of procedure and function declarations are explained in Chapter 6.

Procedure and function subnodes

Procedures and functions may contain their own declaration sections for their arguments and local identifiers. In addition, a procedure or function can be subdivided into logical components which are inserted into the body of that procedure or function, and are stored as execution subnodes. Such execution subnodes allow you to follow a top-down approach in implementing an algorithm without the need to introduce separate procedures to perform every single step. The complete list of permitted subnodes is discussed in Chapter 6.

Attributes

For every node in the model tree you can specify additional information in the form of *attributes*. AIMMS lets you view and change the values of these attributes in an *attribute form* that can be opened for every node in the tree. An example of an attribute form of an identifier node is shown in Figure 4.2. Such an attribute form shows all the attributes that are possible for a particular

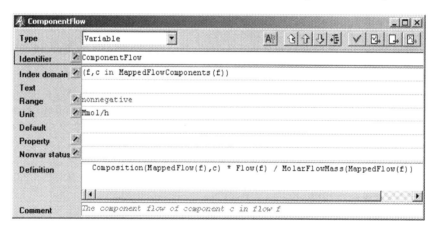

Figure 4.2: Example of an attribute form

node type. For instance, the attribute form of a parameter declaration will show its domain of definition and value range, while the form for a procedure will show the argument list and procedure body. In the attribute form you can enter values that are relevant for your model.

For most attributes in an attribute form AIMMS provides wizards which help you complete the attributes with which you are not familiar. Attribute wizards can be invoked by pressing the small buttons ▓ in front of the attribute fields as shown in Figure 4.2. The wizard dialog boxes may range from presenting a fixed selection of properties, to presenting a relevant subselection of data from your model which can be used to complete the attribute.

Wizards

By providing attribute forms and their associated wizards for the declaration of all identifiers, the amount of syntax knowledge required to set up the model source is drastically reduced. The attribute window of each identifier provides you with a complete overview of all the available attributes for that particular type of identifier. The wizards, in most cases, guide you through one or more dialog boxes in which you can choose from a number of possible options. After selecting the options relevant to your model, AIMMS will subsequently enter these in the attribute form using the correct syntax.

Reduce syntax knowledge

Once your complete model has been compiled successfully, attribute changes to a single identifier usually require only the recompilation of that identifier before the model can be executed again. This local compilation feature of AIMMS allows you to quickly observe the effect of particular attribute changes.

Local compilation

However, when you make changes to some attributes that have global implications for the rest of your model, local compilation will no longer be sufficient. In such a case, AIMMS will automatically recompile the entire model before you can execute it again. Global recompilation is necessary, for instance, when you change the dimension of a particular identifier. In this case global re- compilation is required, since the identifier could be referenced elsewhere in your model.

. . . versus global compilation

The attributes of structuring nodes allow you to specify documentation regarding the contents of that node. You can also provide directives to AIMMS to store a section node and all its offshoots in a separate file which is to be included when the model is compiled. Storing parts of your model in separate model files is discussed in more detail in Section 4.2.

Attributes of structuring nodes

4.2 Creating and managing models

When you begin a new model, AIMMS will automatically create a skeleton model tree suitable for small applications and student assignments. Such a skeleton contains the following nodes:

Creating new models

- a single *declaration section* where you can store the declarations used in your model,

- the predefined procedure *MainInitialization* which is called directly after compiling your model and can be used to initialize your model,
- the predefined procedure *MainExecution* where you can put all the statements necessary to execute the algorithmic part of your application, and
- the predefined procedure *MainTermination* which is called just prior to closing the project.

The model tree also displays the predefined and read-only library module Predeclared Identifiers (see also Section 4.1), which contains all the identifiers predeclared by AIMMS, categorized by function.

Changing the skeleton

Whenever the number of declarations in your model becomes too large to be easily managed within a single declaration section, or whenever you want to divide the execution associated with your application into several procedures, you are free (and advised) to change the skeleton model tree created by AIMMS. You can group particular declarations into separate declaration sections with meaningful names, and introduce your own procedures and functions. You may even decide to remove one or more of the skeleton nodes that are not of use in your application.

Additional structuring of your model

When you feel that particular groups of declarations, procedures and functions belong together in a logical manner, you are encouraged to create a new structuring section with a descriptive name within the model tree, and store the associated model components within it. When your application grows in size, a clear hierarchical structure of all the information stored will help you tremendously in finding your way within your application.

Storage on disk

The contents of a model are stored in one or more binary files with the ".amb" (AIMMS model base) extension. By default the entire model is stored as a single file, but for each book section node ● or module node ● in the tree you can indicate that you want to store the subtree below it in a separate source file. This is especially useful when particular parts of your application are shared with other AIMMS applications, or are developed by other persons. Library modules ● associated with a library project that you have included in your project, are always stored in a separate .amb file.

Separate storage

To store a module or section of your model in a separate source file, open the attribute form of that section node by double-clicking on it in the model explorer. The attribute form of a section is illustrated in Figure 4.3. By selecting the **Write...** command of the **Source file** attribute wizard ▨ on this form, you can select a file where you want all information under the section node to be stored. AIMMS will export the contents of the book section to the indicated file, and enter that file name in the **Source file** attribute of the book section. As a consequence, AIMMS will automatically read the contents of the book section from that file during every subsequent session.

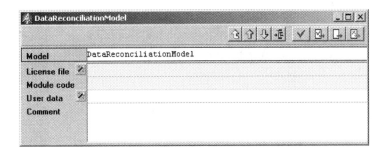

Figure 4.3: Attribute form of a section node

Section 20.2 explains how you can further protect such a .amb file through a VAR license, allowing you to ship it to your customers as an end-user only module. You can create a VAR licensed module using the License file, Module code and User data attributes. These are only visible when your AIMMS license contains a VAR identification code which is unique to you.

Protecting source files

Alternatively, when you are in the **Model Explorer** on the book section node that you want to store in a separate file, you can use the **Edit-Export** menu, to export the contents of the selected section to a separate .amb file. In the latter case, AIMMS will only export a *copy* of the contents of the selected section to the specified .amb file, while the original contents is still stored in the main .amb model file.

Exporting a book section

Likewise, if you want a book section to hold the contents of a section stored in a separate .amb file, you can use the **Read...** command of the **Source file** wizard 🖉. This will let you select an .amb file which will be entered in the **Source file** attribute. As a consequence, the contents of this file will be included into the section during this and any subsequent sessions. Note that any previous contents of a section at the time of entering the **Source file** attribute will be lost completely. By specifying a **Source file** attribute, any changes that you make to the contents of the section after adding a **Source file** attribute will be automatically saved in the corresponding .amb, whenever you save your model.

Adding a book section reference

Alternatively, you can import a copy of the contents of a separate .amb file into your model, by executing the **Edit-Import** menu command on a selected section node in the **Model Explorer**. This will completely replace the current contents of the section with the contents of the .amb file. In this case, however, any changes that you make to the section after importing the .amb file will not be stored in that file, but only in your main model file.

Importing a book section

After each editing session AIMMS will save a backup of the latest version of your model file, and of any module and/or library project included in your project, into the Backup subdirectory of your project directory. Along with this

Backups

automatic backup feature, AIMMS can create additional ASCII files with the ".aim" extension through the **File-Save As** menu in the model explorer. These files contain an ASCII representation of the model and are created for your convenience.

4.2.1 Working with modules and libraries

Name clashes

When you import the contents of a book section node into your model, you may find that particular identifier names in that book section already have been declared in the remainder of your model. If such a name clash occurs, AIMMS will refuse to import the specified .amb file into your model, and present a dialog box indicating which identifiers would cause a name clash when imported.

Avoid name clashes using modules

You can avoid name clashes by using *modules*, which provide their own namespace. Modules allow you to share sections of model source between multiple models, without the risk of running into name clashes. The precise semantics of modules are discussed in full detail in Chapter 31 of the Language Reference.

Creating modules

You can create a module anywhere in your model tree by inserting a *Module* node ⬙ into your tree, as discussed in Section 4.3. For each module you must specify a module prefix through which you can access the identifiers stored in the module. Figure 4.4 illustrates the attributes of a module. If this module

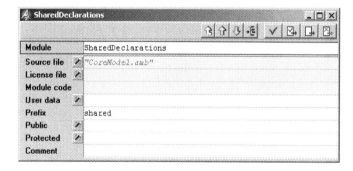

Figure 4.4: The attributes of a *Module* node

contains a parameter GlobalSettings, then outside of the module it can be referenced as shared::GlobalSettings.

AIMMS uses modules to implement those parts of its functionality that can be best expressed in the AIMMS language itself. The available AIMMS system modules include

AIMMS system modules

- a (customizable) implementation of the outer approximation algorithm,
- a scenario generation module for stochastic programming, and
- sets of constants used in the graphical 2D- and 3D-chart objects.

You can include these system modules into your model through the **Settings-System Module...** menu.

If your model becomes too large for a single developer to maintain and develop, you may use *library projects* to create a division of your existing project into sub-projects. The procedure for creating such library projects is discussed in Section 3.1. For each library included in your project, AIMMS creates a separate library module node at the root the **Model Explorer**, as illustrated in Figure 4.5. When creating a new library the associated library module will initially

Library projects ...

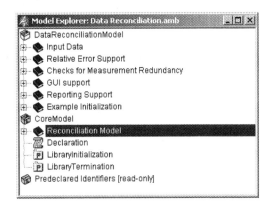

Figure 4.5: A library module containing the core model formulation

be empty. In the library module of Figure 4.5, one section from the original model tree in Figure 4.1 has already been moved into the newly created library.

Contrary to modules, whose principle aim is to let you share a common set of identifier and procedure declarations among multiple models, library projects allow you to truly divide an AIMMS project into subprojects. With every library project you cannot only associate a module in the model tree, but AIMMS lets you also develop pages and menus for the graphical user interface within a library project. Within an AIMMS project that includes such a library project, you can use the model, pages and menus to compose the entire application in a modular way.

... for modular development

Moving identifiers to modules and libraries	When you move identifiers from the main model to a module or a library module, references to such identifiers in the main model may become invalid because because they become part of a different namespace. In accordance with the automatic name change support described in Section 5.2.1, AIMMS will automatically change all references to the identifier in the model source, project pages, and case files to include the module prefix, unless the reference is included in the module or library itself. In occasional situations, however, the automatic name change support of AIMMS may fail to detect such references, for instance, when an identifier name is included in a data initialization statement of a subset of AllIdentifiers.
Library initialization and termination	Each library may provide two procedures *LibraryInitialization* and *LibraryTermination*. If you specify these procedures, they should contain all statements necessary to properly initialize the data associated with a library prior to it first use, and provide the library with a possibility to save its internal state prior to closing a project. The *LibraryInitialization* and *LibraryTermination* procedures are called directly after *MainInitialization* en *MainTermination*, respectively, in the order in which libraries are included into the project.

4.3 Working with trees

Working with trees	The trees used in the various developer tools inside AIMMS offer very similar functionality to the directory tree in the Windows™ Explorer. Therefore, if you are used to working with the Windows Explorer, you should have little difficulty understanding the basic functionality offered by the trees in the AIMMS tools. For novice users, as well as for advanced users who want to understand the differences to the Windows Explorer, this section explains the fine details of working with trees in AIMMS, using the context of the model tree.
Expanding and collapsing branches	Branches in a tree (i.e. intermediate nodes with subnodes) have a small expansion box in front of them containing either a plus or a minus sign. Collapsed branches have a plus sign ⊞, and can be expanded one level by a single click on the plus sign (to show *more* information). Expanded branches have a minus sign ⊟, and can be collapsed by a single click on the minus sign (to show *less* information). Alternatively, a node can be expanded or collapsed by double clicking on its icon. Leaf nodes have no associated expansion box.
Double-clicking a node	When you double-click (or press **Enter**) on the name of any node in a tree, AIMMS will invoke the most commonly used menu command that is specific for each tree.

- In the **Model Explorer**, the double-click is identical to the **Edit-Attributes** menu, which opens the attribute window for the selected node.

■ In the **Identifier Selector**, the double-click is identical to the **Edit-Open With** menu, which opens a view window to simultaneously display the contents of the selection.

■ In the **Page** and **Template Manager**, the double-click is identical to the **Edit-Open** menu, which opens the page or template.

■ In the **Menu Builder**, **Data Manager**, and **Data Management Setup** tool, the double-click is identical to the **Edit-Properties** menu, which opens the appropriate **Properties** dialog box.

Alternatively, you can open the attribute form or **Properties** dialog box of any node type using the **Properties** button on the toolbar.

To create a new node in the model tree you must position the cursor at the node in the tree *after* which you want to insert a new node. You can create a new node here:

Creating new nodes

■ by clicking on one of the node creation icons or on the toolbar

■ by selecting the item **Insert...** from the right-mouse menu, or

■ by pressing the **Ins** key on the keyboard.

The toolbar contains creation icons for the most common node types. You can select the **New...** icon to select further node types.

Once you have clicked the **New...** icon on the toolbar, or selected the **Insert...** menu from the right-mouse menu, or have pressed the **Ins** key, a dialog box as shown in Figure 4.6 appears from which you have to select a node type.

Selecting a node type

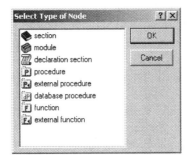

Figure 4.6: Dialog box for selecting a node type

The dialog box shows only those node types that are allowed at the particular position in the tree. You can select a node type by a single mouse click, or by typing in the first letter of the node type that you want to insert. When there are more node types that begin with the same letter (as in Figure 4.6), re-type that letter to alternate over all possibilities.

Naming the node	After you have selected a node type, it is inserted in the model tree, and you have to enter a name for the new node. In the model tree, all node names must consist only of alphanumeric characters and underscores, and must start with a letter. In addition, the names of structuring nodes may contain spaces. For most node types their node names have to be unique throughout the model. The only, quite natural, exception are declaration sections which accept either the predefined name *Declaration* or a name unique throughout the model.
Expanding branches without subnodes	When you want to add subnodes to a branch, you must first expand the branch. If you do not do this, a new node will be inserted directly after the branch, and not as a subnode. Expanding an empty branch will result in an empty subtree being displayed. After expansion you can insert a new node in the usual manner.
Renaming existing nodes	You can rename a selected node by pressing the **F2** button, or single clicking on the node name. After changing the name, press the **Enter** key to action the change, or the **Esc** key to cancel. When the node is an identifier declaration, a procedure, or a function which is used elsewhere in the model (or displayed on a page in the graphical user interface), AIMMS will, if asked, automatically update such references to reflect the name change.
Multiple selections	Unlike the Windows Explorer, AIMMS lets you make multiple selections within a tree which you can delete, cut, copy and paste, or drag and drop. The nodes in a selection do not even have to be within the same branch. By left-clicking in combination with the **Ctrl** key you can add or delete single nodes from the selection. By left-clicking in combination with the **Shift** key you can add all nodes between the current node and the last selected node.
Deleting nodes and branches	You can delete all nodes in a selection by selecting **Delete** from the right-mouse menu, or by pressing the **Del** key. When the selection contains branch nodes, AIMMS will also delete all child nodes contained in that branch.
Cut, copy, paste and duplicate	With the **Cut**, and **Copy** and **Paste** items from the **Edit** menu, or right-mouse menu, you can cut or copy the current selection from the tree, and paste it elsewhere. In addition to the usual way of pasting, which copies information from one position to another, AIMMS also supports the **Paste as Duplicate** operation in the **Identifier Selector**, the **Template Manager** and the **Menu Builder**. This form of pasting makes no copy of the node but only stores a reference to it. In this way changes in one node are also reflected in the other.
Drag and drop support	In addition to the cut, and copy and paste types of operation, you can drag a node selection and drop it onto another position in the model tree, or in any of the other tools offered by AIMMS. Thus you can, for instance, easily move a declaration section to another position in the model tree, or add an identifier

selection to a particular data category in the data manager or to an existing selection in the selection manager.

By pressing the **Shift** or **Ctrl** keys during a drag-and-drop action, you can alter its default action. In combination with the **Shift** key, AIMMS will *move* the selection to the new position, while the **Ctrl** key will *copy* the selection to the new position. With the **Shift** and **Control** key pressed simultaneously, you activate the special *find* function explained in the next paragraph. AIMMS will show the type of action that is performed when you drop the selection by modifying the mouse pointer, or by displaying a stop sign when a particular operation is not permitted.

Copying or moving with drag and drop

AIMMS offers several tools for finding model-related information quickly and easily.

Searching for identifiers

- When the attribute of an identifier, or the body of a procedure or function, contains a reference to another identifier within your application, you can pop up the attribute form of that identifier by simply clicking on the reference and selecting the **Attributes...** item from the right-mouse menu.
- With the **Find...** item from the **Edit** menu (or the **Find** button 🔍 on the toolbar) you can search for all occurrences of an identifier in your entire model or in a particular branch. The **Find** function also offers the possibility of restricting the search to only particular node attributes.
- The **Identifier Selector** offers an advanced tool for creating identifier selections on the basis of one or more dynamic criteria. You can subsequently select a view from the **View Manager** to display and/or change a subset of attributes of all identifiers in the selection simultaneously. Selections and views are discussed in full detail in Chapter 7.
- By dragging a selection of identifiers onto any other tree while pressing the **Ctrl** and **Shift** key simultaneously, AIMMS will highlight those nodes in the tree onto which the selection is dropped, in which the identifiers in the selection play a role. This form of drag and drop support does not only work with identifier selections, but can be used with selections from any other tree as well. Thus, for instance, you can easily find the pages in which a particular identifier is used, or find all pages that use a particular end-user menu or toolbar.

Chapter 5

Identifier Declarations

This chapter

This chapter shows you how to add new identifier declarations using the **Model Explorer** and how to modify existing identifier declarations. The chapter also explains how any changes you make to either the name or the domain of an identifier are propagated throughout the remainder of your model.

5.1 Adding identifier declarations

Identifiers

Identifiers form the heart of your model. All data are stored in identifiers, and the bodies of all functions and procedures consist of statements which compute the values of one identifier based on the data associated with other identifiers.

Adding identifiers

Adding an identifier declaration to your model is as simple as adding a node of the desired type to a global declaration section (or to a declaration section local to a particular procedure or function), as explained in Section 4.3. AIMMS will only allow you to add identifier declarations inside declaration sections.

Identifier types

There are many different types of identifiers. Each identifier type corresponds to a leaf node in the model tree and has its own icon, consisting of a white box containing one or more letters representing the identifier type. When you add an identifier to a declaration section of your model in the model tree, you must first select its identifier type from the dialog box as presented in Figure 5.1.

Identifier name

After you have selected the identifier type, AIMMS adds a node of the specified type to the model tree. Initially, the node name is left empty, and you have to enter a unique identifier name. If you enter a name that is an AIMMS keyword, an identifier predeclared by AIMMS itself, or an existing identifier in your model, AIMMS will warn you of this fact. By pressing the **Esc** key while you are entering the identifier name, the newly created node is removed from the tree.

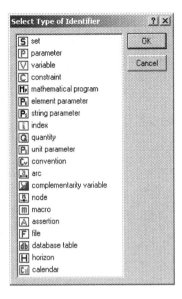

Figure 5.1: Choosing an identifier type

There is no strict limit to the length of an identifier name. Therefore, you are advised to use clear and meaningful names, and not to worry about either word length or the intermingling of small and capital letters. AIMMS offers special features for name completion such as **Ctrl-Spacebar** (see Section 5.2), which allow you to write subsequent statements without having to retype the complete identifier names. Name completion in AIMMS is also case consistent.

Meaningful names are preferable

In addition, when an identifier is multidimensional, you can immediately add the index domain to the identifier name as a parenthesized list of indices that have already been declared in the model tree. Alternatively, you can provide the index domain as a separate attribute of the identifier in its attribute form. Figure 5.2 illustrates the two ways in which you can enter the index domain of an identifier. In both cases the resulting list of indices will appear in the model tree as well as in the **Index Domain** attribute of the attribute form of that identifier. In the **Index Domain** attribute it is possible, however, to provide a further restriction to the domain of definition of the identifier by providing one or more domain conditions (as explained in full detail in the Language Reference). Such conditions will not appear in the model tree.

Index domain

The identifier declarations in the model tree can be used independently of the order in which they have been declared. This allows you to use an identifier anywhere in the tree. This order independence makes it possible to store identifiers where you think they should be stored logically. This is different to most other systems where the order of identifier declarations is dictated by the order in which they are used inside the model description.

Unrestricted order of declarations

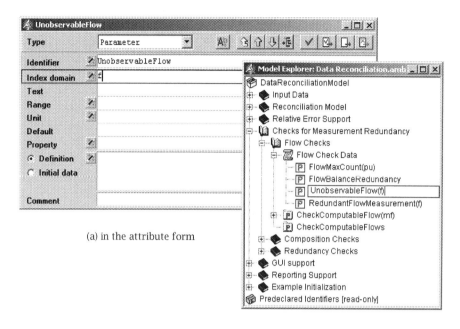

(a) in the attribute form

(b) in the model explorer

Figure 5.2: Specifying an index domain

Identifier scope   In general, all identifiers in an AIMMS model are known globally, unless they have been declared inside a local declaration section of a procedure or function. Such identifiers are only known inside the procedure or function in which they have been declared. When you declare a local identifier with the same name as a global identifier, references to such identifiers in the procedure or function will evaluate using the local rather than the global identifier.

Local declarations   Local identifiers declared in procedures and functions are restricted to particular types of identifier. For example, AIMMS does not allow you to declare constraints as local identifiers in a procedure or function, as these identifier types are always global. Therefore, when you try to add declarations to a declaration section somewhere in the model tree, AIMMS only lists those types of nodes that can be inserted at that position in the model tree.

Declarations via attributes   As an alternative to explicitly adding identifier nodes to the model tree, it is sometimes possible that AIMMS will implicitly define one or more identifiers on the basis of attribute values of other identifiers. The most notable examples are indices and (scalar) element parameters, which are most naturally declared along with the declaration of an index set. These identifiers can, therefore, be specified implicitly via the **Index** and **Parameter** attributes in the attribute form of a set. Implicitly declared identifiers do not appear as separate nodes in the model tree.

5.2 Identifier attributes

The attributes of identifier declarations specify various aspects of the identifier which are relevant during the further execution of the model. Examples are the index domain over which the identifier is declared, its range, or a definition which expresses how the identifier can be uniquely computed from other identifiers. For the precise interpretation of particular attributes of each identifier type, you are referred to the AIMMS Language Reference, which discusses all identifier types in detail.

Identifier attributes

The attributes of an identifier are presented in a standard form. This form lists all the relevant attributes together with the current values of these attributes. The attribute values are always presented in a textual representation, consisting of either a single line or multiple lines depending on the attribute. Figure 5.3 illustrates the attribute form of a variable ComponentFlow(f,c). The

Attribute window

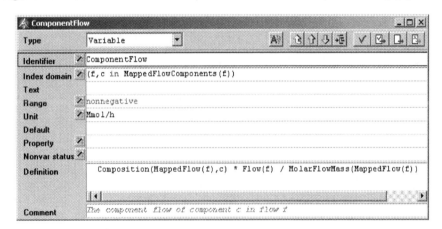

Figure 5.3: Identifier attributes

attributes specify, for instance, that the variable is measured in Mmol/h, and provide a definition in terms of other parameters and variables.

You do not need to enter values for all the attributes in an attribute window. In fact, most of the attributes are optional, or have a default value (which is not shown). You only have to enter an attribute value when you want to alter the behavior of the identifier, or when you want to provide a value that is different to the default.

Default values

*Entering
attribute text ...*

You can freely edit the text of almost every attribute field, using the mechanisms common to any text editor. Of course, you will then need to know the syntax for each attribute. The precise syntax required for each attribute is described in the AIMMS Language Reference book.

*... or using
attribute
wizards*

To help you when filling in attributes, AIMMS offers specialized wizards for most of them. These wizards consists of (a sequence of) dialog boxes, which help you make specific choices, or pick identifier names relevant for specifying the attribute. An example of an attribute wizard is shown is Figure 5.4. In this

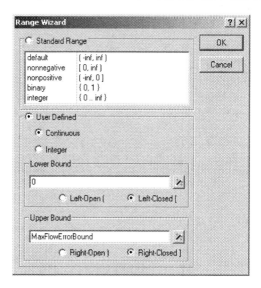

Figure 5.4: Example of an attribute wizard

wizard, the numerical range of a particular parameter or variable is specified as the user-defined interval [0,MaxFlowErrorBound]. After completing the dialog box, the result of filling in the wizard is copied to the attribute window with the correct syntax.

*Mandatory use
of wizards*

Some of the attribute fields are not editable by hand, but require you to always use the associated wizard. AIMMS requires the use of wizards, whenever this is necessary to keep the model in a consistent state. Examples are (non-empty) **Index** and **Parameter** attributes of sets, the **Base unit** attribute of quantities, as well as the VAR licensing attributes of the main model and section nodes.

*Identifier
reference
support*

Even when you decide to enter an attribute into a field manually, AIMMS still offers support to help you enter such a field quickly and easily. If your application contains a large number of identifiers and/or if the names of these identifiers are long, then it may be difficult to remember all the exact names. There are two ways to let AIMMS help you in filling in the appropriate names in an attribute field:

- you can drag and drop the names from the model tree into the field, or
- with the name completion feature you can let AIMMS fill in the remainder of the name based on only the first few characters typed.

When filling in an attribute field, you can drag any identifier node in the model tree to a particular location in the attribute field. As a result, AIMMS will copy the identifier name, with its index domain, at the location where you dropped the identifier.

Dragging identifiers

When you use the **Ctrl-Spacebar** combination anywhere in an attribute field, AIMMS will complete any incomplete identifier name at the current cursor position wherever possible. With the **Ctrl-Shift-Spacebar** combination AIMMS will also complete keywords and predefined procedure and function names. When there are more than one possibilities, a menu of choices is presented as in Figure 5.5. In this menu the first possible extension will be selected and the

Name completion ...

Figure 5.5: Name completion

selection will be updated as you type. When an identifier name is complete, applying name completion will cause AIMMS to extend the identifier by its index domain as specified in its declaration.

By pressing **Ctrl-Spacebar** in a string that contains the :: or . characters, AIMMS will restrict the list of possible choices as follows.

... applied to the :: and . characters

- If the name in front of the :: character is a module or library module prefix, AIMMS will show all the identifiers contained in the module, or all identifiers contained in the interface of the library module, respectively.
- If the string to complete refers to a property in a PROPERTY statement, and the name in front of the . character is an identifier, AIMMS will show all properties available for the identifier (based on its type).
- If the string to complete refers to an option in an OPTION statement, and the string in front of the . character refers to an element of the set AllSolvers, AIMMS will show all options available for that solver.
- In all other cases, if the name in front of the . character is an identifier, AIMMS will show all the suffices available for the identifier (based on its declaration).

5.2.1 Navigation features

Navigation features

From within an attribute window, there are several menus and buttons available to quickly access related information, such as the position in the model tree, identifier attributes and data, and context help on identifier types, attributes and keywords.

Browsing the model tree

From within an attribute window you can navigate further through the model tree by using the navigation buttons displayed at the top of the window.

- The **Parent** 🗁, **Previous** 🗔 and **Next Attribute Window** 🗔 buttons will close the current attribute window, and open the attribute window of the parent, previous or next node in the model, respectively.
- The **Location in Model Tree** 🗔 button will display the model tree and highlight the position of the node associated with the current attribute window.

Viewing identifier details

When an identifier attribute contains a reference to a particular identifier in your model, you may want to review (or maybe even modify) the attributes or current data of that identifier. AIMMS provides various ways to help you find such identifier details:

- by clicking on a particular identifier reference in an identifier attribute, you can open its attributes window through the **Attributes** item in the right-mouse pop-up menu,
- you can locate the identifier declaration in the model tree through the **Location in Model Tree** item in the right-mouse pop-up menu, and
- you can view (or modify) the identifier's data through the **Data** item in the right-mouse pop-up menu (see Section 5.4).

Context help sensitive help

Through either the **Context Help** button 🗔 on the toolbar, or the **Help on** item in the right-mouse pop-up menu, you can get online help for the identifier type, its attributes and keywords used in the attribute fields. It will open the section in one of the AIMMS books or help files, which provides further explanation about the topic for which you requested help.

5.3 Committing attribute changes

Syntax checking

The modifications that you make to the attributes of a declaration are initially only stored locally within the form. Once you take further action, the changes in your model will be checked syntactically and committed to the model. There are three ways to do this.

- **Check and commit** [icon]. This command checks the current values of the attributes for syntax errors, and if there are no errors the new values are applied to the model.
- **Check, commit and close** [icon]. Same as check and commit, but if there are no errors it also closes the current form. Since this is the most frequently used action, you can also invoke it by pressing **Ctrl-Enter**.
- **Commit and close** [icon]. This command does *not* check the current values, but simply applies them to the model and then closes the form. The changes will be checked later, when the entire model is checked or when you re-open and check the form yourself.
- **Discard** [icon]. If you do not want to keep any of the changes you made in the attribute form, you can discard them using the **Discard** button.

In addition to committing the changes in a single attribute form manually as above, the changes that you have made in any attribute form are also committed when you save the model (through the **File-Save** menu), or recompile it in its entirety (through the **Run-Compile All** menu).

Saving the model

It is quite common to rename an existing identifier in a modeling application because you consider that a new name would better express its intention. In such cases, you should be aware of the possible consequences for your application. The following questions are relevant.

Renaming identifiers

- Are there references to the (old) identifier name in other parts of the model?
- Are there case files that contain data with respect to the (old) identifier name?
- Are there pages in the end-user interface that display data with respect to the (old) identifier name?

If the answer to any of these questions is yes, then changing the identifier name could create problems.

AIMMS helps you in dealing with the possible consequences of name changes by offering the following support:

Automatic name changes

- AIMMS updates all references to the identifier throughout the model text, and in addition,
- AIMMS keeps a log of the name change (see also Section 2.5), so that when AIMMS encounters any reference to the old name in either a page or in a case file, the new name will be substituted.

Beware of
structural
changes

Problems arise when you want to change the index domain of an identifier, or remove an identifier, while it is still referenced somewhere in your application. Such changes are called *structural*, and are likely to cause errors in pages and cases. In general, these errors cannot be recovered automatically. To help you locate possible problem areas, AIMMS will mark all pages and cases that contain references to changed or deleted identifiers. To check how a change really affects these pages and cases, you should open them, make any required adaptations to deal with the errors, and resave them.

Modifying
identifier type

You can modify the type of a particular identifier in the model tree via the identifier type drop-down list Element Parameter in the attribute window of the identifier. The drop-down list lets you select from all identifier types that are compatible with the current identifier type. Alternatively, you can change the identifier type via the **Edit-Change Type** menu.

Incompatible
attributes

Before a change of identifier type is actually committed, AIMMS displays the dialog box illustrated in Figure 5.6, which lists all the attributes of the identifier that are not compatible with the newly selected type. If you do not want

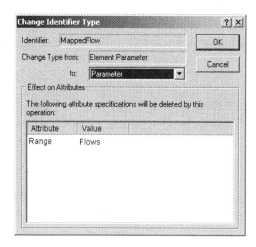

Figure 5.6: The **Change Identifier Type** dialog box

such attributes to be deleted, you should cancel the operation at this point. When you allow AIMMS to actually perform the type change, the incompatible attributes will be deleted.

5.4 Viewing and modifying identifier data

When you are developing your model (or are reviewing certain aspects of it later on), Aimms offers facilities to directly view (and modify) the data associated with a particular identifier. This feature is very convenient when you want to enter data for an identifier during the development of your model, or when you are debugging your model (see also Section 8.1) and want to look at the results of executing a particular procedure or evaluating a particular identifier definition.

Viewing identifier data

Via the **Data** button ![A] available in the attribute window of every global identifier (see, for instance, Figure 5.3), Aimms will pop up one of the *data pages* as illustrated in Figure 5.7. Data pages provide a view of the *current contents* of

The Data button

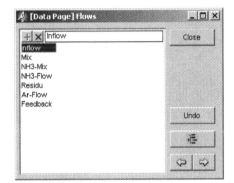
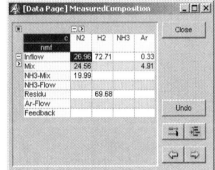

Figure 5.7: Data pages of a set and a 2-dimensional parameter

the selected identifier. Which type of data page is shown by Aimms depends on the type of the identifier. The data page on the left is particular to one-dimensional root sets, while the data page on the right is appropriate for a two-dimensional parameter.

For variables (and similarly for constraints), Aimms will display a pivot table containing all the indices from the index domain of the variable plus one additional dimension containing all the suffices of the variable that contain relevant information regarding the solution of the variable. Depending on the properties set for the variable, this dimension may contain a varying number of suffices containing sensitivity data related to the variable.

Data pages for variables and constraints

Data pages can also be opened directly for a selected identifier node in the model tree using either the **Edit-Data** menu, or the **Data** command in the right-mouse pop-up menu. Additionally, you can open a data page for any identifier referenced in an attribute window by selecting the identifier in the text, and applying the **Data** command from the right-mouse pop-up menu.

Viewing data in the Model Explorer

Multidimen-
sional identifiers

For multidimensional identifiers, AIMMS displays data using a default view which depends on the identifier dimension. Using the 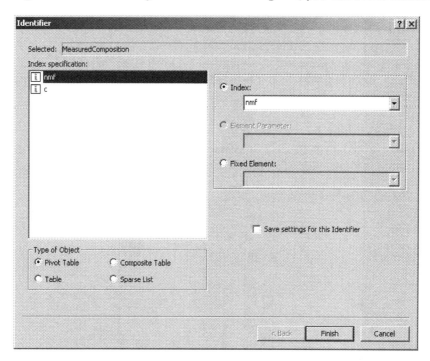 button on the data page you can modify this default view. As a result, AIMMS will display the dialog box illustrated in Figure 5.8. In this dialog box, you can select whether

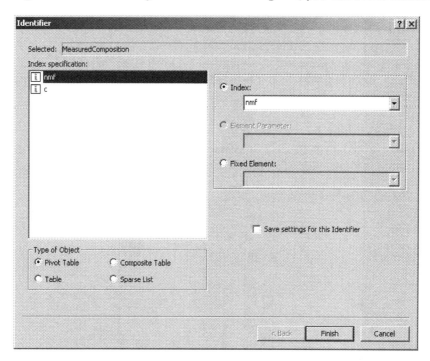

Figure 5.8: Selecting a data page type

you want to view the data in a sparse list object, a composite table object, a pivot table object or in the form of a (rectangular) table. Additionally, you can indicate that you want the view to be *sliced* (see also Section 10.4), by selecting fixed elements for one or more dimensions. For every sliced dimension, AIMMS will automatically add a floating index to the data page, allowing you to view the data for every element in the sliced dimension.

Saving your
choice

If you want to always use the same data page settings for a particular identifier, you can save the choices you made in Figure 5.8. As a result, AIMMS will save the data page as an ordinary end-user page in the special **All Data Pages** section of the **Page Manager** (see also Section 12.1). If you so desire, you can further edit this page, and, for instance, add additional related identifiers to it which will subsequently become visible when you view the identifier data in the **Model Explorer**.

Whenever there is a page in the **All Data Pages** section of the page manager with the fixed name format [Data Page] followed by the name of an identifier of your model, AIMMS will use this page as the data page for that identifier. This enables you to copy a custom end-user page, that you want to use as a data page for one or more identifiers, to the **All Data Pages** section of the page manager, and rename it in the prescribed name format. When you remove a page from the **All Data Pages** section, AIMMS will again open a default data page for that identifier. If you hold down the **Shift** key while opening a data page, AIMMS will always use the default data page.

End-user page as data page

Normally, AIMMS will only allow you to open data pages of global identifiers of your model. However, within the AIMMS debugger (see also Section 8.1), AIMMS also supports data pages for local identifiers within a (debugged) procedure, enabling you to examine the contents of local identifiers during a debug session.

Global and local identifiers

Chapter 6

Procedures and Functions

This chapter

This chapter describes how you can add procedures and functions to a model. It also shows how you can add arguments and local identifiers to procedures and functions. In addition, it illustrates how the body of a procedure or function can be broken down into smaller pieces of execution code, allowing you to implement procedures and functions in a top-down approach.

6.1 Creating procedures and functions

Procedures and functions

Procedures and functions are the main means of executing the sequential tasks in your model that cannot be expressed by simple functional relationships in the form of identifier definitions. Such tasks include importing or exporting your model's data from or to an external data source, the execution of data assignments, and giving AIMMS instructions to optimize a system of simultaneous equations.

Creating procedures and functions

Procedures and functions are added as a special type of node to the model tree, and must be placed in the main model, or in any book section. They cannot be part of declaration sections, which are exclusively used for model identifiers. Procedures and functions are represented by folder icons, which open up when the node is expanded. Figure 6.1 illustrates an example of a procedure node in the model tree.

Naming and arguments

After you have inserted a procedure or function node into the tree, you have to enter its name. If you want to add a procedure or function with arguments, you can add the argument list here as well. Alternatively, you can specify the argument list in the attribute window of the procedure or function. The full details for adding arguments and their declaration as identifiers, local to the procedure or function, are discussed in Section 6.2. Whether or not the arguments are fully visible in the tree is configurable using the **View** menu.

Model Explorer: Data Reconciliation.amb
DataReconciliationModel
Input Data
Reconciliation Model
Relative Error Support
Checks for Measurement Redundancy
Flow Checks
Flow Check Data
CheckComputableFlow(mf)
Declaration
mf
Declaration
CheckComputableFlows
Composition Checks
Redundancy Checks
GUI support
Reporting Support
Example Initialization
Predeclared Identifiers [read-only]

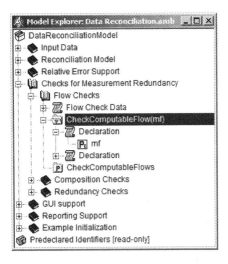

Figure 6.1: Example of a procedure node

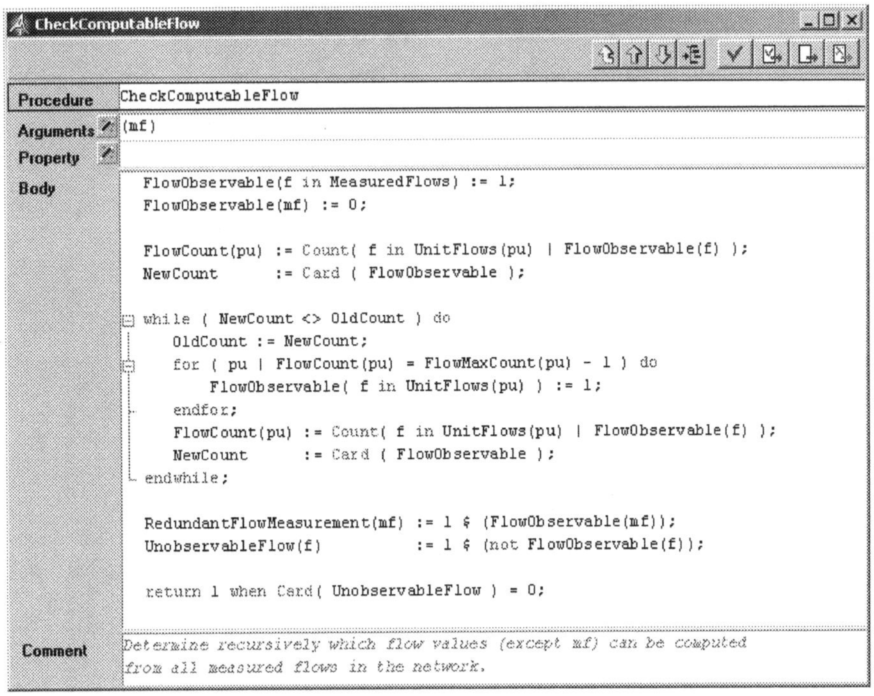

Procedure and function attributes

The attribute window of a procedure or function lets you specify or view aspects such as its list of arguments, the index domain of its result, or the actual body. The body may merely consists of a SOLVE statement to solve an optimization model, but can also consist of a sequence of execution and flow control statements. An example of the attribute window of a procedure node within

CheckComputableFlow
Procedure
CheckComputableFlow
Arguments
(mf)
Property
Body
FlowObservable(f in MeasuredFlows) := 1;
FlowObservable(mf) := 0;
FlowCount(pu) := Count(f in UnitFlows(pu) | FlowObservable(f));
NewCount := Card (FlowObservable);
while (NewCount <> OldCount) do
OldCount := NewCount;
for (pu | FlowCount(pu) = FlowMaxCount(pu) - 1) do
FlowObservable(f in UnitFlows(pu)) := 1;
endfor;
FlowCount(pu) := Count(f in UnitFlows(pu) | FlowObservable(f));
NewCount := Card (FlowObservable);
endwhile;
RedundantFlowMeasurement(mf) := 1 $ (FlowObservable(mf));
UnobservableFlow(f) := 1 $ (not FlowObservable(f));
return 1 when Card(UnobservableFlow) = 0;
Comment
Determine recursively which flow values (except mf) can be computed
from all measured flows in the network.

Figure 6.2: Example of procedure attributes

the model tree is illustrated in Figure 6.2. The contents of the **Body** attribute is application-specific, and is irrelevant to a further understanding of the material in this section.

Specifying function domain and range

When the resulting value of a function is multidimensional, you can specify the index domain and range of the result in the attribute form of the function using the **Index Domain** and **Range** attributes. Inside the function body you can make assignments to the function name as if it were a local (indexed) parameter, with the same dimension as specified in the **Index Domain** attribute. The most recently assigned values are the values that are returned by the function.

6.2   Declaration of arguments and local identifiers

Specifying arguments

All (formal) arguments of a procedure or function must be specified as a parenthesized, comma-separated, list of non-indexed identifier names. All formal arguments must also be declared as local identifiers in a declaration section local to the procedure or function. These local declarations then specify the further domain and range information of the arguments. If an argument has not been declared when you create (or modify) a procedure or function, AIMMS will open the dialog box illustrated in Figure 6.3 which helps you add the appropriate declaration quickly and easily. After completing the dialog box, AIMMS will

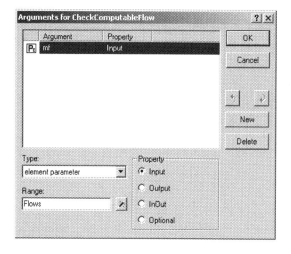

Figure 6.3: **Argument Declaration** dialog box

automatically add a declaration section to the procedure or function, and add the arguments displayed in the dialog box to it, as illustrated in Figure 6.1.

An important aspect of any argument is its input-output type, which can be   *Input-output*
specified by selecting one of the Input, InOut, Output or Optional properties   *type*
in the **Argument Declaration** dialog box. The input- output type determines
whether any data changes you make to the formal arguments are passed back
to the actual arguments on leaving the procedure. The precise semantic meaning of each of these properties is discussed in the AIMMS Language Reference
book.

The choices made in the **Argument Declaration** dialog box are directly re-   *Argument*
flected in the attribute form of the local identifier added to the model tree by   *attributes*
AIMMS. As an example, Figure 6.4 shows the attribute form of the single argument mf of the procedure CheckComputableFlow added in Figure 6.3. In the dialog

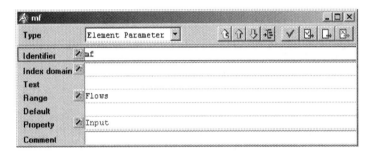

Figure 6.4: Declaration form of a procedure argument

box of Figure 6.3 it is not possible to modify the dimension of a procedure or
function argument directly. If your procedure or function has a multidimensional argument, you can specify this with the **Index Domain** attribute of the
argument after the argument has been added as a local identifier to the model
tree.

For every call to the procedure or function, AIMMS will verify whether the types   *Prototype*
of all the actual arguments match the prototypes supplied for the formal argu-   *checking*
ments, including the supplied index domain and range. For full details about
argument declaration refer to the AIMMS Language Reference book.

In addition to arguments, you can also add other local identifiers to declara-   *Local*
tion sections within procedures and functions. Such local identifiers are only   *declarations*
known inside the function or procedure. They are convenient for storing temporary data that is only useful within the context of the procedure or function,
and have no global meaning or interpretation.

Not all types supported

Not all identifier types can be declared as local identifiers of a procedure or function, because of the global implications they may have for the AIMMS execution engine. When you try to add a local identifier to a procedure or function, AIMMS will only offer those identifier types that are actually supported within a procedure or function. An example of an identifier type that cannot be declared locally is a constraint.

Not all attributes supported

In addition, for local identifiers, AIMMS may only support a subset of the attributes that are supported for global identifiers of the same type. For instance, AIMMS does not allow you to specify a **Definition** attribute for local sets and parameters. In the attribute window of local identifiers such non-supported attributes are automatically removed when you open the associated attribute form.

6.3   Specifying the body

Statements

In the **Body** attribute of a procedure or function you can specify the

- assignments,
- execution statements such as SOLVE or READ/WRITE,
- calls to other procedures or functions in your model, and
- flow control statements such as FOR, WHILE or IF-THEN-ELSE

which perform the actual task or computation for which the procedure or function is intended. The precise syntax of all execution statements is discussed in detail in the AIMMS Language Reference book.

Automatic outlining

When you are constructing a procedure or function whose execution consists of a large number of (nested) statements, it may not always be easy or natural to break up the procedure or function into a number of separate procedures. To help you maintain an overview of such large pieces of execution code, AIMMS will automatically add outlining support for common flow control statement and multiline comments. The minus button ⊟ that appears in front of the statement allows you to collapse the statement to a single line block and the and plus button ⊞ allows you to expand the statement to its full extent again.

Execution blocks

In addition, you can break down the body of a procedure or function into manageable pieces using one or more execution blocks. Any number of AIMMS statements enclosed between BLOCK and ENDBLOCK keywords can be graphically collapsed into a single block. The text in the single line comment following the BLOCK keyword is used as display text for the collapsed block. An example of a procedure body containing two collapsed execution blocks is given in Figure 6.5.

Figure 6.5: Example of a procedure body with execution blocks

When you are entering statements into a body of a procedure or function, AIMMS can help you to add identifier references to the body quickly and easily:

Identifier references

- you can drag and drop the names from the model tree into text
- with the name completion feature you can let AIMMS complete the remainder of the name based on only the first characters typed.

The precise details of drag-and-drop support and name completion of identifiers are discussed in Sections 4.3 and 5.2.

When you are entering the body of a procedure or function, you may want to review the attributes or current data of a particular identifier referenced in the body. AIMMS offers various ways to help you find such identifier details:

Viewing identifier details

- through a text based search in the model tree, you can locate the specific identifier node and open its attribute form (see Section 4.3),
- by clicking on a particular identifier reference in the body, you can open its attributes form through the **Attributes** item in the right-mouse pop-up menu,
- you can locate the identifier declaration in the model tree through the **Location in Model Tree** item in the right-mouse pop-up menu, and
- you can view (or modify) the identifier's data through the **Data** item in the right-mouse pop-up menu (see Section 5.4).

*Viewing
procedure
details*

Similarly, while you are referencing a procedure or function inside the body of another procedure or function, AIMMS will provide prototype information of such a procedure or function as soon as you enter the opening bracket (or when you hover with the mouse pointer over the procedure or function name). This will pop up a window as illustrated in Figure 6.6. This tooltip window

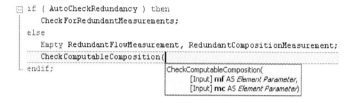

Figure 6.6: Prototype info of a procedure

displays all arguments of the selected procedure or function, their respective data types, as well as their *Input-Output* status. The latter enables you to assess the (global) effect on the actual arguments of a call to the procedure.

6.4   Syntax checking, compilation and execution

*Performing a
syntax check*

Using either **Check and commit** or **Check, commit and close** as discussed in Section 5.3 AIMMS will compile the procedure or function in hand, and point out any syntax error in its body. If you do not want to compile a procedure or function, but still want to commit the changes, you should use the **Commit and close** button. All edits are ignored when you close the window using the **Discard** button.

*Partial
recompilation*

Before executing any procedure in your model, AIMMS will automatically verify whether your model needs to be recompiled, either partially or fully. In most cases, there is no need for AIMMS to recompile the entire model after a modification or addition of a new identifier, a procedure or a function. For instance, when you have only changed the body of a procedure, AIMMS needs only to recompile that particular procedure.

*Complete
recompilation*

However, if you change the index domain of an identifier or the number of arguments of a procedure or function, each reference to such an identifier, procedure or function needs to be verified for correctness and possibly changed. In such cases, AIMMS will (automatically) recompile the entire model before any further execution can take place. Depending on the size of your model, complete recompilation may take some time. Note that either partial or complete recompilation will only retain the data of all identifiers present prior to compilation, to the extent possible (data cannot be retained when, for instance, the dimension of an identifier has changed).

AIMMS supports several methods to initiate procedural model execution. More specifically, you can run procedures

Running a procedure

- from within another procedure of your model,
- from within the graphical user interface by pressing a button, or when changing a particular identifier value, or
- by selecting the **Run procedure** item from the right-mouse menu for any procedure selected in the **Model Explorer**.

The first two methods of running a procedure are applicable to both developers and end-users. Running a procedure from within the **Model Explorer** a useful method for testing the correct operation of a newly added or modified procedure.

Chapter 7

Viewing Identifier Selections

Identifier overviews

Although the **Model Explorer** is a very convenient tool to organize all the information in your model, it does not allow you to obtain a simultaneous overview of a group of identifiers that share certain aspects of your model. By mutual comparison of important attributes (such as the definition), such overviews may help you to further structure and edit the contents of your model, or to discover oversights in a formulation.

This chapter

To assist you in creating overviews that can help you analyze the interrelationships between identifiers in your model, AIMMS offers the **Identifier Selector** tool and **View** windows. This chapter helps you understand how to create meaningful identifier selections with the **Identifier Selector**, and how to display such selections using different views.

7.1 Creating identifier selections

Select by similarity

When you are developing or managing a large and complicated model, you sometimes may need an overview of all identifiers that have some sort of similarity. For example, it may be important to have a simultaneous view of

- all the constraints in a model,
- all variables with a definition,
- all parameters using a certain domain index, or
- all identifiers that cover a specific part of your model.

Identifier selections

In AIMMS, you can create a list of such identifiers using the configurable **Identifier Selector** tool. This tool helps you to create a selection of identifiers according to a set of one or more criteria of varying natures. You can let AIMMS create a once only selection directly in the **Model Explorer**, or create a compound selection in the **Identifier Selector**, which allows you to intersect or unite multiple selections.

If you need a selection only once, then you can create it directly in the **Model Explorer** by   *Creating once only selections*

- either manually selecting one or more nodes in the tree, or
- using the **View-Selection** menu to create a custom selection based on one or more of the conditional selection criteria offered by AIMMS (explained below).

In both cases, the resulting list of selected identifiers will be highlighted in the model tree. If you like, you can narrow down or extend the selection by applying one or more subsequent conditional selections to the existing selection.

If you need a specific selection more than once, then you can create it in the **Identifier Selector** tool. The **Identifier Selector** consists of a tree in which each node contains one of the three types of identifier selectors described below. Figure 7.1 illustrates an example selector tree.   *The Identifier Selector*

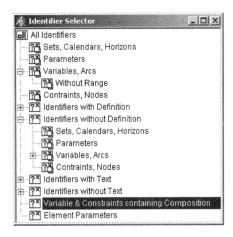

Figure 7.1: The selector tree

In the **Identifier Selector** tool, you can add nodes corresponding to several types of identifier selectors: *Selector types*

- a *node-based selector* ▣, where all the identifiers below one or more user-selected nodes in the model tree are added to the selection,
- a *conditional selector* ▣, where the list of identifiers is created dynamically on identifier type and/or the contents of one of their respective attributes,
- a *set-dependent selector* ▣, where the list of identifiers is created dynamically based on a specific set in either the domain or range of identifiers, or
- a *type-based selector* ▣, where the list of identifiers consists of all variables of a certain type (e.g. free, nonnegative, binary) or all constraints of a certain type ($\leq$, $=$ or $\geq$).

While the above four selectors allow you to define selections based on a symbolic criteria the five types of identifier selectors below allow you to specify selections based on individual criteria. The main purpose of these selector is to define selections that can be used in the Math Program Inspector (see Chapter 9).

- an *element-dependent selector* 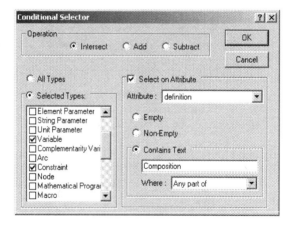, where the list of individual identifiers is created dynamically based of the occurrence of one or more specific elements in the domain,
- a *scale-based selector*, where the list of identifiers is built up from all variables and constraints for which the ratio between the largest absolute value and the smallest absolute value in the corresponding row or column of the matrix exceeds a given value,
- a *status-based selector*, where the list of identifiers is built up from all variables and constraints for which the solution satisfies some property (e.g. feasible, basic, at bound), or
- a *value-based selector*, where the list of identifiers is built up from all variables and constraints for which the level, bound, marginal, or bound violation value satisfy satisfy some property.

Through the **View-Selection** menu in the **Model Explorer** you can only create a new, or refine an existing, selection using a *conditional selector*.

Selection dialog box   To create a selector, AIMMS offers special dialog boxes which let you specify the criteria on which to select. As an example the dialog box for creating a conditional selector is illustrated in Figure 7.2. In it, you can select (by double

Figure 7.2: The **Conditional Selector** dialog box

clicking) one or more identifier types that you want to be part of the selection and filter on specific attributes that should be either empty, nonempty, or should contain a particular string.

The tree structure in the **Identifier Selector** defines combinations of selectors by applying one of the set operators *union*, *difference* or *intersection* with respect to the identifier selection represented by the parent node. The root of the tree always consists of the fixed selection of all model identifiers. For each subsequent child node you have to indicate whether the node should add identifiers to the parent selection, should remove identifiers from the parent selection, or should consider the intersection of the identifiers associated with the current and the parent selection. Thus, you can quickly compose identifier selections that satisfy multiple selection criteria. The type of set operation applied is indicated by the icon of each node in the identifier selector.

Compound selections

In the **Model Explorer**, the *union*, *difference* and *intersection* operations apply to the identifier selection that is currently highlighted in the model tree. You can use them to add identifiers to the current selection, to remove identifiers from the current selection, or filter the current selection by means of an additional criterion.

Refining model tree selections

The list of identifiers that results from a (compound) identifier selector can be used in one of the following ways:

Using selections

- you can display the identifiers in a **View** window of your choice (explained in the next section),
- you can restrict the set of variables and constraints initially displayed in the **Math Program Inspector** (see Chapter 9),
- by dragging and dropping a selector into the **Model Explorer**, the corresponding identifiers will be highlighted in the model tree, or
- you can use the selector in the definition of a *case type* or *data category* in the **Data Management Setup** tool (see Chapter 17).

The drag-and-drop features of AIMMS make it very easy to fill a **View** window with identifiers from either the model tree, the **Identifier Selector** or other **View** windows. If you drag-and-drop a selection into any other AIMMS window, AIMMS will interpret this as a special search action to highlight all occurrences of the selected identifiers as follows:

Advanced drag and drop

- in the *model tree* all identifiers in the selection will be highlighted,
- in the *page* or *template tree* all pages that contain reference to the identifiers in the selection will be highlighted,
- in an end-user *page*, in edit mode, all objects that contain references to the identifiers will be selected,
- in the *menu builder tree*, AIMMS will highlight all menu items that reference one or more identifiers in the selection, and
- in the *data management setup tree*, all data categories and case types that contain one or more identifiers in the selection will be highlighted.

In addition, AIMMS also supports the 'drag-and-drop-search' action in a **View** window by pressing both the **Shift** and **Control** key during the drop operation.

7.2   Viewing identifier selections

Overview of attributes

After you have created an identifier selection, in either the **Model Explorer** or in the **Identifier Selector**, you may want to compare or simultaneously edit multiple attributes of the identifiers in the selection. In general, sequential or simultaneous, opening of all the corresponding single attribute forms is impractical or unacceptable for such a task. To assist, AIMMS offers special identifier **View** windows.

Identifier views

A **View** window allows you to view one or more attributes simultaneously for a number of identifiers. Such a **View** window is presented in the form of a table, where each row represents a single identifier and each column corresponds to a specific attribute. The first column is always reserved for the identifier name. An example of an identifier **View** window is given in Figure 7.3.

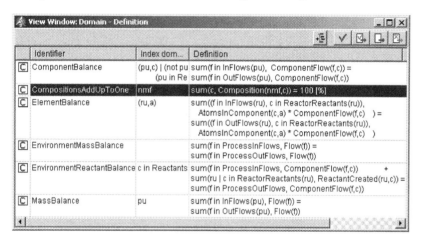

Figure 7.3: Example of a **View** window

Editing in a View window

In addition to simply viewing the identifier content in a **View** window, you can also use it to edit individual entries. To edit a particular attribute of an identifier you can just click on the relevant position in the **View** window and modify the attribute value. This can be convenient, for instance, when you want to add descriptive text to all identifiers for which no text has yet been provided, or when you want to make consistent changes to units for a particular selection of identifiers. As in a single attribute form, the changes that you make are not committed in the model source until you use one of the special compile buttons at the top right of the window (see also Section 5.3).

Using the **Edit-Open with** menu, or the **Open with** item in the right- mouse pop-up menu, you can open a particular **View** window for any identifier selection in the model explorer or in the identifier selector. Selecting the **Open with** menu will open the **View Manager** dialog box as displayed in Figure 7.4. In the **View Manager** you must select one of the available *view window defini-*

Opening a View window

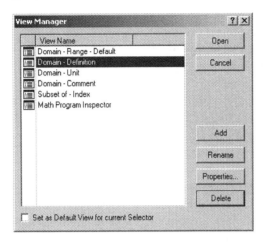

Figure 7.4: The **View Manager** dialog box

tions, with which to view the given identifier selection. For every new project, the **View Manager** will automatically contain a number of basic view window definitions that can be used to display the most common combinations of identifier attributes.

Using the **Add**, **Delete** and **Properties** buttons in the **View Manager**, you can add or delete view window definitions to the list of available definitions, or modify the contents of existing definitions. For every view window definition that you add to the list or want to modify, AIMMS will open the **View Definition Properties** dialog box as illustrated in Figure 7.5. With this dialog box you can

Creating a view window definition

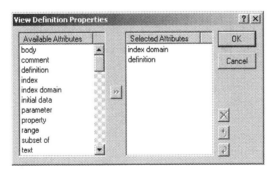

Figure 7.5: **View Definition Properties** dialog box

add or remove attributes from the list of attributes that will be shown in the

View window, or change the order in which the particular attributes are shown.

Changing the View window contents

After opening a **View** window, with the contents of a particular identifier selection, you can add new identifiers to it by dragging and dropping other identifier selections from either the **Model Explorer** or the **Identifier Selector**. Using the **Edit-Delete** menu or the **Del** key, on the other hand, you can delete any subselection of identifiers from the **View** window. At any time you can save the modified identifier selection as a new node in the identifier selector tree through the **View-Selection-Save** menu.

Selecting identifier groups

Besides selecting individual identifiers from the model tree, you can also select whole groups of identifiers by selecting their parent node. For example, if you drag-and-drop an entire declaration section into a **View** window, all the identifiers contained in that section will be added to the view.

Specifying a default view

As can be seen at the bottom of the **View Manager** dialog box in Figure 7.4, it is possible to associate a default view definition with every selector in the **Identifier Selector**. As a consequence, whenever you double-click on such an identifier selector node, AIMMS will immediately open a default **View** window with the current contents of that selection.

Chapter 8

Debugging and Profiling an AIMMS Model

After you have developed an (optimization) model in AIMMS, it will most probably contain some unnoticed logical and/or programming errors. These errors can cause infeasible solutions or results that are not entirely what you expected. Also, you may find that the execution times of some procedures in your model are unacceptably high for their intended purpose, quite often as the result of only a few inefficiently formulated statements. To help you isolate and resolve such problems, AIMMS offers a number of diagnostic tools, such as a debugger and a profiler, which will be discussed in this chapter.

This chapter

8.1 The AIMMS debugger

When your model contains logical errors or programming errors, finding the exact location of the offending identifier declarations and/or statements may not be easy. In general, incorrect results might be caused by:

Tracking modeling errors

- incorrectly specified attributes for one or more identifiers declared in your model (most notably in the INDEX DOMAIN and DEFINITION attributes),
- logical oversights or programming errors in the formulation of one or more (assignment) statements in the procedures of your model,
- logical oversights or programming errors in the declaration of the variables and constraints comprising a mathematical program, and
- data errors in the parametric data used in the formulation of a mathematical program.

If the error is in the formulation or input data of a mathematical program, the main route for tracking down such problems is the use of the **Math Program Inspector** discussed in Chapter 9. Using the **Math Program Inspector** you can inspect the properties of custom selections of individual constraints and/or variables of a mathematical program.

Errors in mathematical programs

To help you track down errors that are the result of misformulations in assignment statements or in the definitions of defined parameters in your model, AIMMS provides a *source debugger*. You can activate the AIMMS debugger through the **Tools-Diagnostic Tools-Debugger** menu. This will add a **Debug-**

The AIMMS debugger

ger menu to the system menu bar, and, in addition, add the **Debugger** toolbar illustrated in Figure 8.1 to the toolbar area. You can stop the AIMMS debugger

Figure 8.1: The **Debugger** toolbar

through the **Debugger-Exit Debugger** menu.

Debugger functionality

Using the AIMMS debugger, you can

- set conditional and unconditional breakpoints on a statement within the body of any procedure or function of your model, as well as on the evaluation of set and parameter definitions,
- step through the execution of procedures, functions and definitions, and
- observe the effect of single statements and definitions on the data within your model, either through tooltips within the observed definitions and procedure bodies, or through separate data pages (see also Section 5.4)

Setting breakpoints in the body

Within the AIMMS debugger you can set breakpoints on any statement in a procedure or function body (or on the definition of a defined set, parameter or variable) by selecting the corresponding source line in the body of the procedure or function, and choosing the **Debugger-Breakpoints-Insert/Remove** menu (or the **Insert/Remove Breakpoint** button on the **Debugger** toolbar). After you have set a breakpoint, this is made visible by means of red dot in the left margin of selected source line, as illustrated in Figure 8.2.

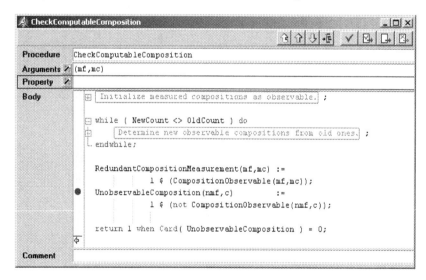

Figure 8.2: Setting a breakpoint in a procedure body

Alternatively, you can set a breakpoint on a procedure, function or on a defined set, parameter or variable by selecting the corresponding node in the **Model Explorer**, and choosing the **Debugger-Breakpoints-Insert/Remove** menu. As a result, AIMMS will add a breakpoint to the first statement contained in the body of the selected procedure or function. The name of a node of any procedure, function or defined set, parameter or variable with a breakpoint is displayed in red in the model tree, as illustrated in Figure 8.3.

Setting breakpoints in the model tree

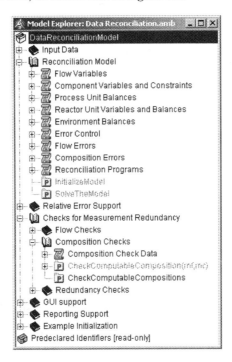

Figure 8.3: Viewing procedures with breakpoints in the **Model Explorer**

Once you have set a breakpoint in your model, AIMMS will automatically stop at this breakpoint whenever a line of execution arrives at the corresponding statement. This can be the result of

Entering the debugger

- explicitly running a procedure within the **Model Explorer**,
- pushing a button on an end-user page which results in the execution of one or more procedures, or
- opening an end-user (or data) page, which requires the evaluation of a defined set or parameter.

Whenever the execution stops at a breakpoint, AIMMS will open the corresponding procedure body (or the declaration form of the defined set, parameter or variable), and show the current line of execution through the breakpoint pointer ⇨, as illustrated in Figure 8.4.

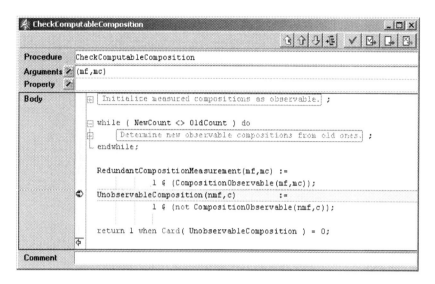

Figure 8.4: Arriving at a breakpoint

Interrupting execution

Even when you have not set breakpoints, you can still enter the debugger by explicitly *interrupting* the current line of execution through the **Run-Stop** menu (or through the **Ctrl-Shift-S** shortcut key). It will pop up the **Stop Run** dialog box illustrated in Figure 8.5 When you have activated the AIMMS debugger prior

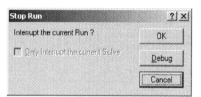

Figure 8.5: The **Stop Run** dialog box

to execution, the **Debug** button on it will be enabled, and AIMMS will enter the debugger when you push it. By pushing the **OK** or **Cancel** button, AIMMS will completely stop or just continue executing, respectively.

Stepping through statements

Once AIMMS has interrupted a line of execution and entered the debugger, you can step through individual statements by using the various step buttons on the **Debugger** toolbar and follow the further flow of execution, or observe the effect on the data of your model. AIMMS offers several methods to step through your code:

- the **Step Over** ☐ method runs a single statement, and, when this statement is a procedure call, executes this in its entirety,
- the **Step Into** ☐ method runs a single statement, but, when this statement is a procedure call, sets the breakpoint pointer to the first statement in this procedure,

- the **Step Out** method runs to the end of the current procedure and sets the breakpoint pointer to the statement directly following the procedure call in the calling context, and
- the **Run To Cursor** method runs in a single step from the current position of the breakpoint pointer to the current location of the cursor, which should be *within the current procedure.*

In addition, AIMMS offers some methods to continue or halt the execution:

- the **Continue Execution** method continues execution, but will stop at any breakpoint it will encounter during this execution,
- the **Finish Execution** method finishes the current line of execution, ignoring any breakpoints encountered,
- the **Halt** method immediately halts the current line of execution.

Whenever you are in the debugger, AIMMS allows you to interactively examine the data associated with the identifiers in your model, and observe the effect of statements in your source code. The most straightforward method is by simply moving the mouse pointer over a reference to an identifier (or identifier *slice*) within the source code of your model. As a result, AIMMS will provide an overview of the data contained in that identifier (slice) in the form of a tooltip, as illustrated in Figure 8.6. The tooltip will provide global information about

Examining identifier data

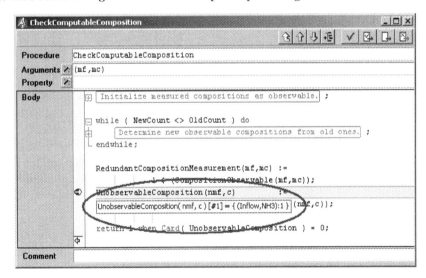

Figure 8.6: Observing the current data of an identifier through a tooltip

the identifier slice at hand, such as

- its name and indices,
- the number of elements or non-default data values (in brackets), and
- the first few elements or non-default data value in the form of a list consisting of tuples and their corresponding values.

Detailed identifier data

If you need to examine the effect of a statement on the data of a particular identifier in more detail, you can simply open a **Data Page**, as described in Section 5.4, or observe the effect on ordinary end-user pages. Within a debugger session, AIMMS supports data pages for both global and local identifiers, thereby allowing you to examine the contents of local identifiers as well. After each step in the debugger AIMMS will automatically update the data on any open end-user or data page.

Viewing the call stack

Whenever you are in the debugger, the **Call Stack** button 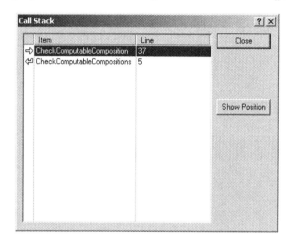 on the **Debugger** toolbar will display the **Call Stack** dialog box illustrated in Figure 8.7. With it

Figure 8.7: The **Call Stack** dialog box

you get a detailed overview of the stack of procedure calls associated with the current line of execution. It enables you to observe the flow of execution at the level of procedures associated with the current position of the breakpoint pointer. After selecting a procedure or definition in the **Call Stack** dialog box, the **Show Position** button will open its attribute window at the indicated line.

Viewing and modifying breakpoints

After you have inserted a number of breakpoints into your model, you can get an overview of all breakpoints through the **Show All Breakpoints** button ▦ This button will invoke the **List of Breakpoints** dialog box illustrated in Figure 8.8. For each breakpoint, AIMMS will indicate whether it is enabled or disabled (i.e. to be ignored by the AIMMS debugger). Through the buttons on the right hand side of the dialog box you can

- disable breakpoints,
- enable previously disabled breakpoints,
- delete breakpoints, and
- create new breakpoints.

Alternatively, you can disable or remove all breakpoints simultaneously using the **Disable All Breakpoints** button ▦ and the **Remove All Breakpoints**

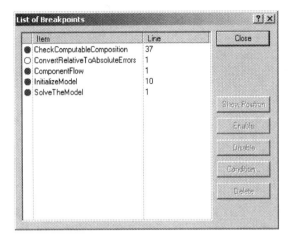

Figure 8.8: The **List of Breakpoints** dialog box

button ![]

In addition, by pushing the **Condition** button on the **List of Breakpoints** dialog box, you can add a condition to an existing breakpoint. It will open the
Breakpoint Condition dialog box illustrated in Figure 8.9. The condition must

Conditional
breakpoints

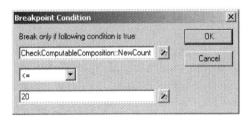

Figure 8.9: The **Breakpoint Condition** dialog box

consist of a simple numerical, element or string comparison. This simple comparison can only involve scalar identifiers, identifier slices or constants. Free
indices in an identifier slice are only allowed when they are fixed within the
breakpoint context (e.g. through a for loop). AIMMS will only stop at a conditional breakpoint, when the condition that you have specified is met during a
particular call. Conditional breakpoints are very convenient when, for instance,
a procedure is called very frequently, but only appears to contain an error in
one particular situation which can be detected through a simple comparison.

8.2   The AIMMS profiler

Once your model is functionally complete, you may find that the overall computational time requirement set for the application is not met. If your applica-

Meeting time
requirements
with solvers

tion contains optimization, and most of the time is spent by the solver, finding a remedy for the observed long solution times may not be easy. In general, it involves finding a reformulation of the mathematical program which is more suitable to the selected solver. Finding such a reformulation may require a considerable amount of expertise in the area of optimization.

Meeting time requirements with data execution

It could also be, however, that optimization (if any) only consumes a small part of the total execution time. In that case, the time required for executing the application is caused by data manipulation statements. If total execution time is unacceptably high, it could be caused by inefficiently formulated statements. Such statements force AIMMS to fall back to *dense* instead of *sparse* execution. Chapters 12 and 13 of the Language Reference discuss the principles of the sparse execution engine used by AIMMS, and describe several common pitfalls together with reformulations to remedy them.

The AIMMS *profiler*

AIMMS offers a profiler to help you resolve computational time related issues. The AIMMS profiler enables you to locate the most time-consuming evaluations of

- procedures and functions,
- individual statements within procedures and functions,
- defined sets and parameters, and
- constraints and defined variables during matrix generation.

Activating the profiler

You can activate the AIMMS profiler by selecting the **Tools-Diagnostic Tools-Profiler** menu, which will add a **Profiler** menu to the default system menu bar. If the debugger is still active at this time, it will be automatically deactivated, as both tools cannot be used simultaneously.

Gathering timing information

As soon as you have activated the profiler, AIMMS will start gathering timing information during every subsequent procedure run or definition evaluation, regardless whether these are initiated by pushing a button on an end-user page, by executing a procedure from within the **Model Explorer**, or even by means of a call to the AIMMS API from within an external DLL.

Viewing profiler results

After you have gathered timing information about your modeling application by executing the relevant parts of your application at least once, you can get an overview of the timing results through the **Profiler-Results Overview** menu. This will open the **Profiler Results Overview** dialog box illustrated in Figure 8.10. In it, you will find a list of all procedures that have been executed and identifier definitions that have been evaluated since the profiler was activated.

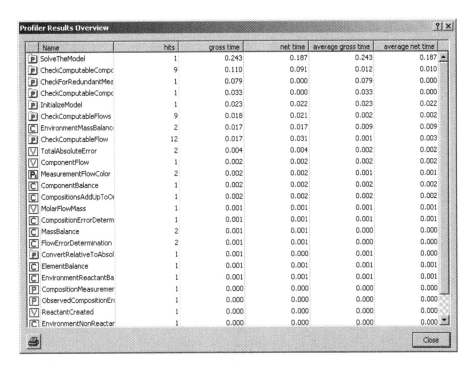

Name	hits	gross time	net time	average gross time	average net time
SolveTheModel	1	0.243	0.187	0.243	0.187
CheckComputableCompc	9	0.110	0.091	0.012	0.010
CheckForRedundantMea	1	0.079	0.000	0.079	0.000
CheckComputableCompc	1	0.033	0.000	0.033	0.000
InitializeModel	1	0.023	0.022	0.023	0.022
CheckComputableFlows	9	0.018	0.021	0.002	0.002
EnvironmentMassBalanc	2	0.017	0.017	0.009	0.009
CheckComputableFlow	12	0.017	0.031	0.001	0.003
TotalAbsoluteError	2	0.004	0.004	0.002	0.002
ComponentFlow	1	0.002	0.002	0.002	0.002
MeasurementFlowColor	2	0.002	0.002	0.001	0.001
ComponentBalance	1	0.002	0.002	0.002	0.002
CompositionsAddUpToOr	1	0.002	0.002	0.002	0.002
MolarFlowMass	1	0.001	0.001	0.001	0.001
CompositionErrorDeterm	1	0.001	0.001	0.001	0.001
MassBalance	2	0.001	0.001	0.000	0.000
FlowErrorDetermination	2	0.001	0.001	0.000	0.000
ConvertRelativeToAbsol	1	0.001	0.000	0.001	0.000
ElementBalance	1	0.001	0.001	0.001	0.001
EnvironmentReactantBa	1	0.001	0.001	0.001	0.001
CompositionMeasuremer	1	0.000	0.000	0.000	0.000
ObservedCompositionEr	1	0.000	0.000	0.000	0.000
ReactantCreated	1	0.000	0.000	0.000	0.000
EnvironmentNonReactar	1	0.000	0.000	0.000	0.000

Figure 8.10: The **Profiler Results Overview** dialog box

For each procedure, function, or defined identifier listed in the **Profiler Results Overview** dialog box, AIMMS will provide the following information:

Detailed timing information

- the number of hits (i.e. the number of times a procedure has been executed or a definition has been evaluated),
- the total gross time (explained below) spent during all hits,
- the total net time (explained below) spent during all hits,
- the average gross time spent during each separate hit, and
- the average net time spent during each separate hit.

The term *gross time* refers to the total time spent in a procedure *including* the time spent in procedure calls or definition evaluations within the profiled procedure. The term *net time* refers to the total time spent *excluding* the time spent in procedure calls or definition evaluations within the profiled procedure.

Gross versus net time

With this timing information you can try to locate the procedures and identifier definitions which are most likely to benefit from a reformulation to improve efficiency. To help you locate these procedures and definitions, the list of procedures and definitions in the **Profiler Results Overview** dialog box can be sorted with respect to all its columns. The most likely criterion for this is to sort by decreasing net time or average net time, which will identify those

Locating time-consuming procedures

procedures and identifier definitions which take up the most time by themselves, either in total or for each individual call. You can open the attribute form of any identifier in the **Profiler Results Overview** dialog box by simply double-clicking on the corresponding line.

Locating offending statements

When you have located a time-consuming procedure, you can can open its attribute form and try to locate the offending statement(s). Whenever the profiler has been activated, AIMMS will add additional profiling columns to the body of a procedure, as illustrated in Figure 8.11. Similarly, AIMMS will add these

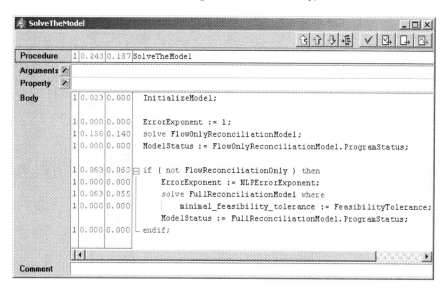

Figure 8.11: Profiling information in an attribute form

profiling columns to the definition attributes of defined identifiers.

Profiling column information

For each statement in the body of a procedure, AIMMS can display various types of profiling data in the profiling columns of an attribute form. As you can see next, this information is even more extensive than for procedures as a whole. The following information is available:

- the number of hits (i.e. the number of times a particular statement has been executed),
- the total gross time spent during all hits,
- the total net time spent during all hits,
- the average gross time spent during each separate hit,
- the average net time spent during each separate hit,
- the actual number of expression evaluations,
- the maximum number of expression evaluations, and
- the sparsity index (i.e. actual number of evaluations as percentage of maximum number of evaluations).

In the context of a procedure body, the difference between gross and net time need not always refer only to the time spent in other procedures (as in the **Profiler Results Overview** dialog box). For selected statements both numbers may have a somewhat different, yet meaningful, interpretation. The list of exceptions is:

Gross versus net time for particular statements

- in flow control statements such as the IF, WHILE and FOR statement (see also Section 8.3 of the Language Reference), the net time refers to the time required to evaluate the statement itself (for instance, its condition) whereas the gross time refers to the time required to execute the entire statement,
- in the SOLVE statement (see also Section 15.3 of the Language Reference), the net time refers to the time spent in the solver, while the gross time refers to the time spent in the solver plus the time required to generate the model.

The actual and maximum expression evaluation count and the sparsity index, may give you a valuable clue why a particular statement takes a long time to execute. The key observation in reducing execution times is that you should avoid the *dense* execution of statements as much as possible, especially when the domain of execution is high-dimensional. You will find detailed information about sparse versus dense execution, as well as many useful hints how to avoid dense execution, in the Chapters 12 and 13 of the Language Reference.

Counting expression evaluations

The expression evaluation count forms a good measure for detecting dense execution. They provide the following information.

Interpretation of evaluation counts

- The *actual expression evaluation count* is the number of expression evaluations as counted by AIMMS during all evaluations of a particular statement. This number can depend on the cardinalities of all sets and multi-dimensional identifiers involved in the statement.
- The *maximum expression evaluation count* is the number of actual expression evaluations that would be necessary if the statement were to be executed in a dense manner.
- The *sparsity index* displays the actual evaluation count as a percentage of the maximum evaluation count.

You can use the following strategy to locate and reformulate time-consuming, densely executed, statements.

Locating densely executed statements

- Locate those statements in a time-consuming procedure that take a disproportional amount of time.
- If the maximum expression evaluation count and the sparsity index are high, then you have very likely found a high-dimensional statement that is executed in a completely or nearly dense manner.

> ■ Search in Chapter 13 whether the statement contains a construct which forces AIMMS to execute this statement in a dense manner and follow the associated instructions to reformulate the construct.
>
> ■ If you cannot find the reason for the disproportional execution time, contact Paragon for their expert advise.

Profiler tooltips

In addition to observing the profiling times in the **Profiler Results Overview** and through the profiling columns in attribute windows, AIMMS also provides profiling tooltips in both the **Model Explorer** and in the attribute windows of procedures and defined identifiers, as long as the profiler is active. An example of a profiling tooltip is given in Figure 8.12. Profiling tooltips can provide a con-

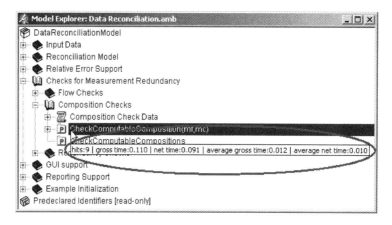

Figure 8.12: Observing profiling information through a tooltip

venient method to quickly observe the profiling information without requiring any further interaction with AIMMS. If you do not want AIMMS to display profiling tooltips while moving your mouse through either the **Model Explorer** or procedure bodies, you can disable them through the **Profiler Setup** dialog box described below, by unchecking the **Show Profiler Values** check mark (see also Figure 8.13).

Profiler listing

If you are interested in a profiling overview comprising your entire modeling application, you can get this through the **Profiler-Create Listing File** menu. This will create a common source listing file of your model text extended with profiling information wherever this is available. Through the **Profiler Setup** dialog box described below you can determine which profiler information will be added to the profiler listing.

For every new project, AIMMS uses a set of default settings to determine which profiling information is displayed in the various available methods to display profiling information. You can modify these settings through the **Profiler-Setup** menu, which will open the **Profiler Setup** dialog box illustrated in Figure 8.13. In this dialog box you can, on a project-wide basis, determine

Setting up profiling columns

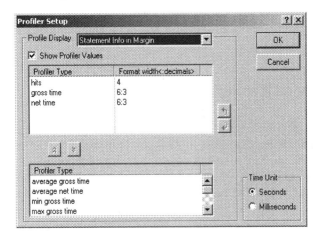

Figure 8.13: The **Profiler Setup** dialog box

- which of the profiling display methods described are enabled (through the **Show Profiler Values** check mark), and
- per such display method, which profiling information is to be displayed, their order, and their corresponding display format.

The settings selected in the **Profiler Setup** dialog box are saved along within the project file, and will be restored when you reopen the project in another AIMMS session.

Through the **Profiler-Pause** menu, you can temporarily halt the gathering of profiling information by AIMMS, while the existing profiling information will be retained. You can use this menu, for example, when you only want to profile the core computational procedures of your modeling application and do not want the profiling information to be cluttered with profiling information that is the result of procedure calls and definition evaluations in the end-user interface. You can resume the gathering of profiling information through the **Profiler- Continue** menu.

Pausing and continuing the profiler

With the **Profiler-Reset** menu, you can completely reset all profiling counters to zero. You can use this menu, if the profiling information of your application has become cluttered. For instance, some procedures may have been executed multiple times and, thus, disturb the typical profiling times required by your entire application. After resetting the profiling counters, you can continue to

Resetting the profiler

gather new profiling information which will then be displayed in the various profiling displays.

Exiting the profiler

You can completely disable the AIMMS profiler through the **Profiler-Exit Profiler** menu. As a result, the gathering of profiling information will be completely discontinued, and all profiling counters will be reset to zero. Thus, when you restart the profiler, all profiling information of a previous session will be lost.

8.3   Observing identifier cardinalities

Observing identifier cardinalities

Another possible cause of performance problems is when one or more multi-dimensional identifiers in your model have missing or incorrectly specified domain conditions. As a result, AIMMS could store far too much data for such identifiers. In addition, computations with these identifiers may consume a disproportional amount of time.

Locating cardinality problems

To help you locate identifiers with missing or incorrectly specified domain conditions, AIMMS offers the **Identifier Cardinalities** dialog box illustrated in Figure 8.14. You can invoke it through the **Tools-Diagnostic Tools-Identifier Cardinalities** menu.

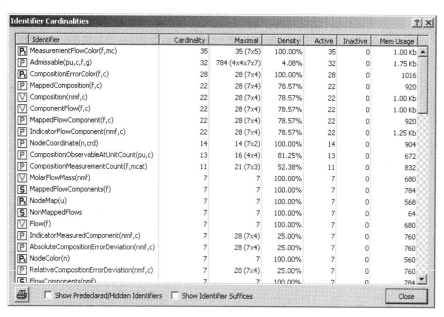

Figure 8.14: The **Identifier Cardinalities** dialog box

The **Identifier Cardinalities** dialog box displays the following information for each identifier in your model:

- the *cardinality* of the identifier, i.e. the total number of non-default values currently stored,
- the *maximal cardinality*, i.e. the cardinality if all values would assume a non-default value,
- the *density*, i.e. the cardinality as a percentage of the maximal cardinality,
- the number of *active* values, i.e. of elements that lie within the domain of the identifier,
- the number of *inactive* values, i.e. of elements that lie outside of the domain of the identifier, and
- the *memory usage* of the identifier, i.e. the amount of memory needed to store the identifier data.

Available information

The list of identifier cardinalities can be sorted with respect to any of these values.

You can locate potential dense data storage problems by sorting all identifiers by their cardinality. Identifiers with a very high cardinality and a high density can indicate a missing or incorrectly specified domain condition. In most real-world applications, the higher-dimensional identifiers usually have relatively few tuples, as only a very small number of combinations have a meaningful interpretation.

Locating dense data storage

If your model contains one or more identifiers that appear to demonstrate dense data storage, it is often possible to symbolicly describe the appropriate domain of allowed tuple combinations. Adding such a condition can be helpful to increase the performance of your model by reducing both the memory usage and execution times.

Resolving dense storage

Another type of problem that you can locate with the **Identifier Cardinalities** dialog box, is the occurrence of *inactive* data in your model. Inactive data can be caused by

Locating inactive data

- the removal of elements in one or more domain sets, or
- data modifications in the identifier(s) involved in the domain restriction of the identifier.

In principle, inactive data does not directly influence the behavior of your model, as the AIMMS execution engine itself will never consider inactive data. Inactive data, however, can cause unexpected problems in some specific situations.

Problems with inactive data

- One of the areas where you have to be aware about inactive data is in the AIMMS API (see also Chapter 30 of the Language Reference), where you have to decide whether or not you want the AIMMS API to pass inactive data to an external application or DLL.

■ Also, when inactive data becomes active again, the previous values are retained, which may or may not be what you intended.

As a result, the occurrence of inactive data in the **Identifier Cardinalities** dialog box may make you rethink its consequences, and may cause you to add statements to your model to remove the inactive data explicitly (see also Section 21.3 of the Language Reference).

Chapter 9

The Math Program Inspector

In this chapter you will find the description of an extensive facility to analyze both the input and output of a generated linear optimization model. This mathematical program inspector allows you to make custom selections of constraints and variables. For each such selection, you can inspect statistics of the corresponding matrix and the corresponding solution. The main purpose of the math program inspector, however, is to assist you in finding causes of infeasibility, unboundedness and unrealistic solutions.

This chapter

The design and contents of the math program inspector is strongly influenced by the papers and contributions of Bruce A. McCarl on misbehaving mathematical programs. The example in the last section of this chapter is a direct reference to his work.

Acknowledgement

9.1 Introduction and motivation

Even though you have taken the utmost care in constructing your linear optimization model, there are often unforeseen surprises that force you to take a further look at the particular generated model at hand. Why is the model infeasible or unbounded? Why is the objective function value so much different from what you were expecting? Why is the number of individual constraints so large? Why are the observed shadow prices so unrealistically high (or low)? These and several other related questions about the matrix and the solution need further investigation.

Unforeseen surprises...

The answer to many model validation questions is not easily discovered, especially when the underlying optimization model has a large number of individual constraints and variables. The amount of information to be examined is often daunting, and an answer to a question usually requires extensive analysis involving several steps. The functionality of the math program inspector is designed to facilitate such analysis.

... are not easily explained

Some of the causes

There are many causes of unforeseen surprises that have been observed in practice. Several are related to the values in the matrix. Matrix input coefficients may be incorrect due to a wrong sign, a typing error, an incorrect unit of measurement, or a calculation flaw. Bounds on variables may have been omitted unintentionally. Other causes are related to structural information. The direction of a constraint may be accidentally misspecified. The subsets of constraints and variables may contain incorrect elements causing either missing blocks of constraints and variables, or unwanted blocks. Even if the blocks are the correct ones, their index domain restrictions may be missing or incorrect. As a result, the model may contain unwanted and/or missing constraints and/or variables.

Purpose

The purpose of the mathematical program inspector included in AIMMS is to provide you with

- insight into the (structure of the) generated model and its solution (if present), and
- a collection of tools to help you discover errors in your model model,

9.2 Functional overview

This section

In this section you will find a description of the functionality available in the mathematical program inspector. Successively, you will learn

- the basics of the trees and windows available in the mathematical program inspector,
- how you can manipulate the contents of the variable and constraint trees through variable and constraint properties, but also using the **Identifier Selector** tool,
- how to inspect the contents and properties of the matrix and solution corresponding to your mathematical program, and
- which analysis you can perform using the mathematical program inspector when your mathematical program is infeasible.

9.2.1 Tree view basics

Viewing generated variables and constraints

The math program inspector window displays the set of all generated variables and all generated constraints, each in a separate tree (see the left portion of Figure 9.1). In these trees, the symbolic identifiers are the first-level nodes and on every subsequent level in the tree, one or more indices are fixed. As a result, the individual variables and constraints in your model appear as leaf nodes in the two tree view windows.

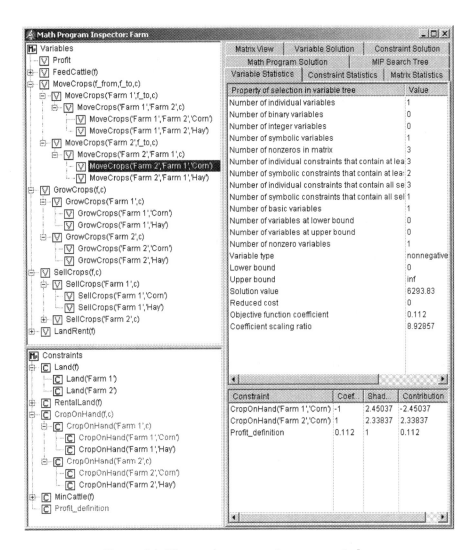

Figure 9.1: The math program inspector window

The math program inspector contains several tabs (see the right portion of Figure 9.1) that retrieve information regarding the selection that has been made in the tree views. Common Windows controls are available to select a subset of variables and constraints (mouse click possibly in combination with the *Shift* or *Control* key). Whenever you select a slice (i.e. an intermediate node in the tree) all variables or constraints in that subtree are selected implicitly. You can use the **Next Leaf Node** and **Previous Leaf Node** buttons in the toolbar for navigational purposes. In Figure 9.1 a single variable has been selected in the variable tree.

Tree view selections

Bookmarks allow you to temporarily tag one or more variables or constraints. While navigating through the tree you will always change the current selection,

Bookmarks

while the bookmarked nodes will not be affected. Whenever you bookmark a node, all its child nodes plus parent nodes are also bookmarked. Using the **Bookmarks** menu you can easily select all bookmarked nodes or bookmark all selected nodes. Bookmarks appear in blue text. Figure 9.1 contains a constraint tree with three bookmarked constraints plus their three parent nodes. You can use the **Next Bookmark** and **Previous Bookmark** buttons in the toolbar for navigational purposes. In case your Figure 9.1 is not displayed in color, the light-gray print indicates the bookmarked selection.

Domain index display order

By default only one index is fixed at every level of the tree views, and the indices are fixed from the left to the right. However, you can override the default index order as well as the subtree depth by using the **Variable Property** or **Constraint Property** dialog on the first-level nodes in the tree. The subtree depth is determined by the number of distinct index groups that you have specified in the dialog.

Finding associated variables/ constraints

The linkage between variables and constraints in your model is determined through the individual matrix coefficients. To find all variables that play a role in a particular constraint selection, you can use the **Associated Variables** command to bookmark the corresponding variables. Similarly, the **Associated Constraints** command can be used to find all constraints that play a role in a particular variable selection. In Figure 9.1, the associated constraint selection for the selected variable has been bookmarked in the constraint tree.

9.2.2 Advanced tree manipulation

Variable and constraint properties

Using the right-mouse popup menu you can access the **Variable Properties** and **Constraint Properties**. On the dialog box you can specify

- the domain index display order (already discussed above), and
- the role of the selected symbolic variable or constraint during infeasibility and unboundedness analysis.

Variable and constraint statistics

The math program inspector tool has two tabs to retrieve statistics on the current variable and constraint selection. In case the selection consists of a single variable or constraint, all coefficients in the corresponding column or row are also listed. You can easily access the variable and constraint statistics tabs by double-clicking in the variable or constraint tree. Figure 9.1 shows the variable statistics for the selected variable.

Popup menu commands

In addition to **Variable Properties** and **Constraint Properties**, you can use the right-mouse popup menu to

- open the attribute form containing the declaration of an identifier,
- open a data page displaying the data of the selected slice,

- make a variable or constraint at the first level of the tree inactive (i.e. to exclude the variable or constraint from the generated matrix during a re-solve), and
- bookmark or remove the bookmark of nodes in the selected slice.

Using the identifier selector you can make sophisticated selections in the variable and/or constraint tree. Several new selector types have been introduced to help you investigate your mathematical program. These new selector types are as follows.

Interaction with identifier selector

- **element-dependency selector:** The element-dependency selector allows you to select all individual variables or constraints for which one of the indices has been fixed to a certain element.
- **scale selector:** The scale selector allows you to find individual rows or columns in the generated matrix that may be badly scaled. The selection coefficient for a row or column introduced for this purpose has been defined as

$$\frac{\text{largest absolute (nonzero) coefficient}}{\text{smallest absolute (nonzero) coefficient}}.$$

The **Properties** dialog associated with the scale selector offers you several possibilities to control the determination of the above selection coefficient.

- **status selector:** Using the status selector you can quickly select all variables or constraints that are either basic, feasible or at bound.
- **value selector:** The value selector allows you to select all variables or constraints for which the value (or marginal value) satisfies some simple numerical condition.
- **type selector:** With the type selector you can easily filter on variable type (e.g. continuous, binary, nonnegative) or constraint type (e.g. less-than-or-equal, equal, greater-than-or-equal).

9.2.3 Inspecting matrix information

Most of the statistics that are displayed on the Variable Statistics tab are self-explanatory. Only two cases need additional explanation. In case a single symbolic (first-level node) has been selected, the *Index domain density* statistic will display the number of actually generated variables or constraints as a percentage of the full domain (i.e. the domain without any domain condition applied). In case a single variable (a leaf node) has been selected, the statistics will be extended with some specific information about the variable such as bound values and solution values.

Variable Statistics tab

Column
coefficients

In case a single variable x_j has been selected, the lower part of the information retrieved through the Variable Statistics tab will contain a list with all coefficients a_{ij} of the corresponding rows i, together with the appropriate shadow prices y_i (see Figure 9.1). The last column of this table will contain the contributions $a_{ij}y_j$ that together with the objective function coefficient c_j make up the reduced cost $\bar{c}_j$ according to the following formula.

$$\bar{c}_j = c_j - \sum_i a_{ij} y_i$$

Constraint
Statistics tab

The Constraints Statistics tab and the Variable Statistics tab retrieve similar statistics. Figure 9.2 shows the constraint statistic for the selection consisting of a single constraint. Note that in this particular case the symbolic form of the constraint definition will also be displayed.

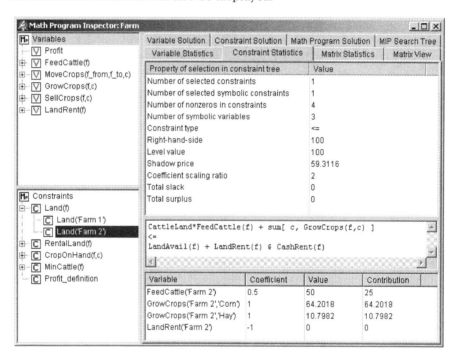

Figure 9.2: The math program inspector window

Row coefficients

In case a single row j has been selected, the lower part of the Constraint Statistics tab will contain all coefficients a_{ij} in the corresponding columns j, together with their level values x_j. The last column of this table lists the contributions $a_{ij}x_j$ that together with the right-hand- side make up either the slack or surplus that is associated with the constraint according to the following formula.

$$\text{slack}_i - \text{surplus}_i = \text{rhs}_i - \sum_j a_{ij} x_j$$

The Matrix Statistics tabs retrieves information that reflects both the selection in the variable tree and the selection in the constraint tree. Among these statistics are several statistical moments that might help you to locate data outliers (in terms of size) in a particular part of the matrix.

Matrix Statistics tab

The Matrix View tab contains a graphical representation of the generated matrix. This view is available in two modes that are accessible through the right-mouse popup menu. The symbolic block view displays at most one block for every combination of symbolic variables and symbolic constraints. The individual block view allows you to zoom in on the symbolic view and displays a block for every nonzero coefficient in the matrix. It is interesting to note that the order in which the symbolic and individual variables and constraints are displayed in the block view follows the order in which they appear in the trees.

Matrix View tab

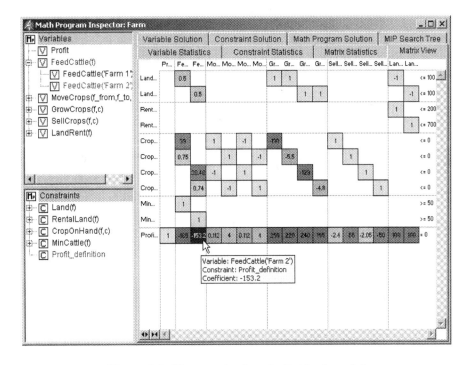

Figure 9.3: The matrix view (individual mode)

The colors of the displayed blocks correspond to the value of the coefficient. The colors will vary between green and red indicating small and large values. Any number with absolute value equal to one will be colored green. Any number for which the absolute value of the logarithm of the absolute value exceeds the logarithm of some threshold value will be colored red. By default, the threshold is set to 1000, meaning that all nonzeros $x \in (-\infty, -1000] \cup$

Block coloring

$[-\frac{1}{1000}, \frac{1}{1000}] \cup [1000, \infty)$ will be colored red. All numbers in between will be colored with a gradient color in the spectrum between green and red.

AIMMS option

The value of the threshold mentioned in the previous paragraph is available as an AIMMS option with name bad_scaling_threshold and can be found in the **Project - Math program inspector** category in the **AIMMS Options** dialog box.

Block tooltips

While holding the mouse inside a block, a tooltip will appear displaying the corresponding variables and constraints. In the symbolic view the tooltip will also contain the number of nonzeros that appear in the selected block. In the individual view the actual value of the corresponding coefficient is displayed.

Block view features

Having selected a block in the block view you can use the right-mouse popup menu to synchronize the trees with the selected block. As a result, the current bookmarks will be erased and the corresponding selection in the trees will be bookmarked. Double-clicking on a block in symbolic mode will zoom in and display the selected block in individual mode. Double-clicking on a block in individual mode will center the display around the mouse.

9.2.4 Inspecting solution information

Solution tabs

The tabs discussed so far are available as long as the math program has been generated. As soon as a solution is available, the next three tabs reveal more details about this solution.

Variable Solution tab

The Variable Solution tab shows the following five columns

- Variable Name,
- Value (i.e. solution/level value),
- Marginal (i.e. reduced cost),
- Basis (i.e. *Basic* or *Nonbasic*), and
- Bound (i.e. *At bound* or *In between bounds*).

By clicking in the header of a column you can sort the table according to that specific column.

Constraint Solution tab

A similar view is available for the constraints in your mathematical program. The Constraint Solution tab contains the following five columns

- Constraint Name,
- Value (i.e. solution),
- Marginal (i.e. shadow price),
- Basis (i.e. *Basic* or *Nonbasic*), and
- Bound (i.e. *Binding* or *Nonbinding*).

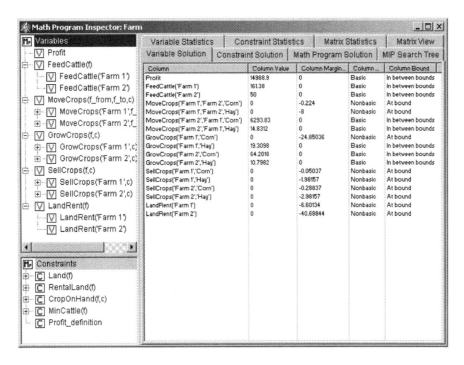

Figure 9.4: The variable solution

The Math Program Solution tab retrieves solution information about the mathematical program that has been solved. This information is similar to that in the AIMMS **Progress** window.

Math Program Solution tab

The lower part of the information retrieved by this tab is used to display logging messages resulting from the **Bound Analysis** and **Unreferenced Identifiers** commands in the **Actions** menu.

Logging messages

Whenever your linear model is a mixed-integer model, the solver will most probably use a tree search algorithm to solve your problem. During the tree search the algorithm will encounter one or more solutions if the model is integer feasible. Once the search is completed, the optimal solution has been found.

Solving MIP models

With the MIP Search Tree tab you can retrieve branching information about the search tree. Currently CPLEX 9 is the only MIP solver in AIMMS that provides this information.

MIP Search Tree tab

The size and shape of the search tree might give you some indication that you could improve the performance of the solver by tuning one or more solver options. Consider the case in which the search tree algorithm spends a considerable amount of time in parts of the tree that do not seem interesting in

Improving the search process

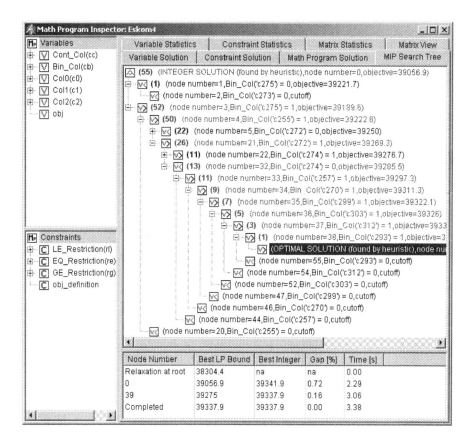

Figure 9.5: The MIP search tree

retrospect. You might consider to use priorities or another branching rule, in an attempt to direct the search algorithm to a different part of the tree in an earlier stage of the algorithm.

Controlling search tree memory usage

Because all structural and statistical information is kept in memory, displaying the MIP search tree for large MIPs might not be a good idea. Therefore, you are able to control to the display (and size) of the search through the options Show_branch_and_bound_tree and Maximum_number_of_nodes_in_tree.

Search tree display

For every node several solution statistics are available. They are the sequence number, the branch type, the branching variable, the value of the LP relaxation, and the value of the incumbent solution when the node was evaluated. To help you locate the integer solutions in the tree, integer nodes and their parent nodes are displayed in blue.

Incumbent progress

The lower part of the MIP Search Tree tab retrieves all incumbent solutions that have been found during the search algorithm. From this view you are able to conclude for example how much time the algorithm spend before finding the optimal solution, and how much time it took to proof optimality.

9.2.5 Performing analysis to find causes of problems

One of the causes of a faulty model may be that you forgot to include one or more variables or constraints in the specification of your mathematical model. The math program inspector helps you in identifying some typical omissions. By choosing the **Unreferenced Identifiers** command (from the **Actions** menu) AIMMS helps you to identify

Unreferenced identifiers

- constraints that are not included in the constraint set of your math program while they contain a reference to one of the variables in the variable set,
- variables that are not included in the variable set of your math program while a reference to these variables does exist in some of the constraints, and
- defined variables that are not included in the constraint set of your math program.

The results of this action are visible through the Math program solution tab.

In some situations it is possible to determine that a math program is infeasible or that some of the constraints are redundant even before the math program is solved. The bound analysis below supports such investigation.

A priori bound analysis

For each constraint with a left-hand side of the form

Implied constraint bounds

$$\sum_j a_{ij} x_j$$

the minimum level value $\underline{b_i}$ and maximum level value $\overline{b_i}$ can be computed by using the bounds on the variables as follows.

$$\underline{b_i} = \sum_{j|a_{ij}>0} a_{ij}\underline{x_j} + \sum_{j|a_{ij}<0} a_{ij}\overline{x_j}$$

$$\overline{b_i} = \sum_{j|a_{ij}>0} a_{ij}\overline{x_j} + \sum_{j|a_{ij}<0} a_{ij}\underline{x_j}$$

By choosing the **Bound Analysis** command (from the **Actions** menu) the above implied bounds are used not only to detect infeasibilities and redundancies, but also to tighten actual right-hand-sides of the constraints. The results of this analysis can be inspected through the Math Program Solution tab. This same command is also used to perform the variable bound analysis described below.

Performing bound analysis

Implied variable bounds . . .

Once one or more constraints can be tightened, it is worthwhile to check whether the variable bounds can be improved. An efficient approach to compute implied variable bounds has been proposed by Gondzio, and is presented without derivation in the next two paragraphs.

. . . for $\leq$ constraints

For i in the set of constraints of the form $\sum_j a_{ij}x_j \leq b_i$, the variable bounds can be tightened as follows.

$$x_k \leq \underline{x_k} + \min_{i|a_{ik}>0} \frac{b_i - \underline{b_i}}{a_{ik}}$$

$$x_k \geq \overline{x_k} + \max_{i|a_{ik}<0} \frac{b_i - \underline{b_i}}{a_{ik}}$$

. . . and $\geq$ constraints

For i in the set of constraints of the form $\sum_j a_{ij}x_j \geq b_i$, the variable bounds can be tightened as follows.

$$x_k \leq \underline{x_k} + \min_{i|a_{ik}<0} \frac{b_i - \overline{b_i}}{a_{ik}}$$

$$x_k \geq \overline{x_k} + \max_{i|a_{ik}>0} \frac{b_i - \overline{b_i}}{a_{ik}}$$

Phase 1 analysis

In case infeasibility cannot be determined a priori (e.g. using the bound analysis described above), the solver will conclude infeasibility during the solution process and return a phase 1 solution. Inspecting the phase 1 solution might indicate some causes of the infeasibility.

Currently infeasible constraints

The collection of currently infeasible constraints are determined by evaluating all constraints in the model using the solution that has been returned by the solver. The currently infeasible constraints will be bookmarked in the constraint tree after choosing the **Infeasible Constraints** command from the **Actions** menu.

Substructure causing infeasibility

To find that part of the model that is responsible for the infeasibility, the use of slack variables is proposed. By default, the math program inspector will add slacks to all variable and constraint bounds with the exception of

- variables that have a definition,
- zero variable bounds, and
- bounds on binary variables.

Adapting the use of slack variables

The last two exceptions in the above list usually refer to bounds that cannot be relaxed with a meaningful interpretation. However these two exceptions can be overruled at the symbolic level through the Analysis Configuration tab of the **Properties** dialog. These properties can be specified for each node at the first level in the tree. Of course, by not allowing slack variables on all variable and constraint bounds in the model, it is still possible that the infeasibility will not be resolved.

Note that to add slacks to variable bounds, the original simple bounds are removed and (ranged) constraints are added to the problem definition.

Slack on variable bounds

$$\underline{x_j} \le x_j + s_j^+ - s_j^- \le \overline{x_j}$$

After adding slack variables as described above, this adapted version of the model is referred to as the elastic model.

Elastic model

When looking for the substructure that causes infeasibility, the sum of all slack variables is minimized. All variables and constraints that have positive slack in the optimal solution of this elastic model, form the substructure causing the infeasibility. This substructure will be bookmarked in the variable and constraint tree.

Minimizing feasibility violations

Another possibility to investigate infeasibility is to focus on a so-called *irreducible inconsistent system* (IIS). An IIS is a subset of all constraints and variable bounds that contains an infeasibility. As soon as at least one of the constraints or variable bounds in the IIS is removed, that particular infeasibility is resolved.

Irreducible Inconsistent System (IIS)

Several algorithms exist to find an *irreducible inconsistent system* (IIS) in an infeasible math program. The algorithm that is used by the AIMMS math program inspector is discussed in Chinneck ([Ch91]). While executing this algorithm, the math program inspector

Finding an IIS

1. solves an elastic model,
2. initializes the IIS to all variables and constraints, and then
3. applies a combination of *sensitivity* and *deletion* filters.

Deletion filtering loops over all constraints and checks for every constraint whether removing this constraint also solves the infeasibility. If so, the constraint contributes to the infeasibility and is part of the IIS. Otherwise, the constraint is not part of the IIS. The deletion filtering algorithm is quite expensive, because it requires a model to be solved for every constraint in the model.

Deletion filtering

The sensitivity filter provides a way to quickly eliminate several constraints and variables from the IIS by a simple scan of the solution of the elastic model. Any nonbasic constraint or variable with zero shadow price or reduced cost can be eliminated since they do not contribute to the objective, i.e. the infeasibility. However, the leftover set of variables and constraint is not guaranteed to be an IIS and deletion filtering is still required.

Sensitivity filtering

Combined filtering

The filter implemented in the math program inspector combines the deletion and sensitivity filter in the following way. During the application of a deletion filter, a sensitivity filter is applied in the case the model with one constraint removed is infeasible. By using the sensitivity filter, the number of iterations in the deletion filter is reduced.

Substructure causing unboundedness

When the underlying math program is not infeasible but unbounded instead, the math program inspector follows a straightforward procedure. First, all infinite variable bounds are replaced by a big constant M. Then the resulting model is solved, and all variables that are equal to this big M are bookmarked as being the *substructure causing unboundedness*. In addition, all variables that have an extremely large value (compared to the expected order of magnitude) are also bookmarked. Any constraint that contains at least two of the bookmarked variables will also be bookmarked.

Options

When trying to determine the cause of an infeasibility or unboundedness, you can tune the underlying algorithms through the following options.

- In case infeasibility is encountered in the presolve phase of the algorithm, you are advised to turn off the presolver. When the presolver is disabled, solution information for the phase 1 model is passed to the math program inspector.
- During determination of the substructure causing unboundedness or infeasibility and during determination of an IIS, the original problem is pertubated. After the substructure or IIS has been found, AIMMS will restore the original problem. By default, however, the solution that is displayed is the solution of the (last) pertubated problem. Using the option Restore_original_solution_after_analysis you can force a resolve after the analysis has been carried out.
- To be able to view the MIP tree that is constructed during the branch-and-bound algorithm, you need CPLEX 9.0 or higher. Besides this, the following option has to be set (before the solve).
    - the option Show_branch_and_bound_tree has to be set to *on*,

9.3 A worked example

This section

The example in this section is adapted from McCarl ([Mc98]), and is meant to demonstrate the tools that were discussed in the previous sections. The example model is used to illustrate the detection of infeasibility and unboundedness. In addition, the example is used to find the cause of an unrealistic solution.

9.3.1 Model formulation

The model considers a collection of farms. For each of these farms several decisions have to be made. These decisions are

A Farm planning model

- the amount of cattle to keep,
- the amount of land to grow a particular crop,
- the amount of additional land to rent, and
- the amount of inter-farm crop transport.

The objective of the farm model is to maximize a social welfare function, which is modeled as the total profit over all farms.

The following notation is used to describe the symbolic farm planning model.

Notation

Indices:

$f, \hat{f}$	*farms*
c	*crops*

Parameters:

C_{fc}^{g}	*unit cost of growing crop c on farm f*
C_{f}^{c}	*unit cost of keeping cattle on farm f*
C_{c}^{m}	*unit transport cost for moving crop c*
C_{f}^{r}	*rental price for one unit of land on farm f*
P_{fc}^{s}	*unit profit of selling crop c grown on farm f*
P_{f}^{c}	*unit profit of selling cattle from farm f*
L_{f}	*amount of land available on farm f*
Q	*amount of land needed to keep one unit of cattle*
Y_{fc}	*crop yield per unit land for crop c on farm f*
D_{fc}	*consumption of crop c by one unit of cattle on farm f*
M_{f}^{c}	*minimum required amount of cattle on farm f*
M_{f}^{r}	*maximum amount of land to be rented on farm f*

Variables:

p	*total profit*
c_{f}	*amount of cattle on farm f*
$m_{f\hat{f}c}$	*amount of crop c moved from farm f to farm $\hat{f}$*
g_{fc}	*amount of land used to grow crop c on farm f*
s_{fc}	*amount of crop c sold by farm f*
r_{f}	*amount of extra land rented by farm f*

*Land
requirement*

The *land requirement* constraint makes sure that the total amount of land needed to keep cattle and to grow crops does not exceed the amount of available land (including rented land).

$$Qc_f + \sum_c g_{fc} \le L_f + r_f, \qquad \forall f$$

*Upper bound on
rental*

The total amount of rented land on a farm cannot exceed its maximum.

$$r_f \le M_f^r, \qquad \forall f$$

Crop-on-hand

The *crop-on-hand* constraint is a crop balance. The total amount of crop exported, crop sold, and crop that has been used to feed the cattle cannot exceed the total amount of crop produced and crop imported.

$$\sum_{\hat{f}} m_{f\hat{f}c} + s_{fc} + D_{fc}c_f \le Y_{fc}g_{fc} + \sum_{\hat{f}} m_{\hat{f}fc}, \qquad \forall (f,c)$$

*Cattle
requirement*

The cattle requirement constraint ensures that every farm keeps at least a prespecified amount of cattle.

$$c_f \ge M_f^c, \qquad \forall f$$

Profit definition

The total profit is defined as the net profit from selling crops, minus crop transport cost, minus rental fees, plus the net profit of selling cattle.

$$p = \sum_f \left(\sum_c \left(P_{fc}^s s_{fc} - C^g g_{fc} - \sum_{\hat{f}} C^m m_{f\hat{f}c} \right) - C_f^r r_f + (P_f^c - C_f^c) c_f \right)$$

*The generated
problem*

Once the above farm model is solved, the math program inspector will display the variable and constraint tree plus the matrix block view as illustrated in Figure 9.3. The solution of the particular farm model instance has already been presented in Figure 9.4.

9.3.2   Investigating infeasibility

*Introducing an
infeasibility*

In this section the math program inspector will be used to investigate an artificial infeasibility that is introduced into the example model instance. This infeasibility is introduced by increasing the land requirement for cattle from 0.5 to 10.

By selecting the **Infeasible Constraints** command from the **Actions** menu, all violated constraints as well as all variables that do not satisfy their bound conditions, are bookmarked. Note, that the solution values used to identify the infeasible constraints and variables are the values returned by the solver after infeasibility has been concluded. The exact results of this command may depend on the particular solver and the particular choice of solution method (e.g. primal simplex or dual simplex).

Locating infeasible constraints

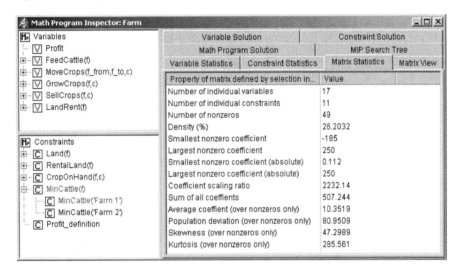

Figure 9.6: An identified substructure causing infeasibility

By selecting the **Substructure Causing Infeasibility** command from the **Actions** menu a single constraint is bookmarked. In this example, one artificial violation variable could not be reduced to zero by the solver used, which resulted in a single infeasibility. Figure 9.6 indicates that this infeasibility can be resolved by changing the right-hand-side of the 'MinCattle' constraint for 'Farm 1'. A closer investigation shows that when the minimum requirement on cattle on 'Farm 1' is decreased from 50 to 30, the infeasibility is resolved. This makes sense, because one way to resolve the increased land requirement for cattle is to lower the requirements for cattle.

Substructure causing infeasibility

By selecting the **Irreducible Inconsistent System** command from the **Actions** menu, an IIS is identified that consists of the three constraints 'RentalLand', 'Land' and 'MinCattle', all for 'Farm 1' (see Figure 9.7).

Locating an IIS

The above IIS provides us with three possible model changes that together should resolve the infeasibility. These changes are

Resolving the infeasibility

1. increase the availability of land for 'Farm 1',
2. change the land requirement for cattle on 'Farm 1', and/or

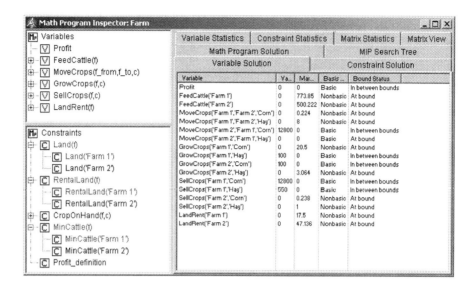

Figure 9.7: An identified IIS

3. decrease the minimum requirement on cattle on 'Farm 1'.

It is up to the producer of the model instance, to judge which changes are appropriate.

9.3.3   Investigating unboundedness

Introducing unboundedness

The example model is turned into an unbounded model by dropping the constraints on maximum rented land, and at the same time, by multiplying the price of cattle on 'Farm 1' by a factor 100 (representing a unit error). As a result, it will become infinitely profitable for 'Farm 1' to rent extra land to keep cattle.

Substructure causing unboundedness

By selecting the **Substructure Causing Unboundedness** command from the **Actions** menu four individual variables are bookmarked, and all of them are related to 'Farm 1'. Together with all constraints that contain two or more bookmarked variables these bookmarked variables form the problem structure that is subject to closer investigation. From the optimal solution of the auxiliary model it becomes clear that the 'FeedCattle' variable, the two 'GrowCrops' variables and the 'LandRent' variables tend to get very large, as illustrated in Figure 9.8.

Resolving the unboundedness

Resolving the unboundedness requires you to determine whether any of the variables in the problem structure should be given a finite bounds. In this case, specifying an upper bound on the 'RentalLand' variable for 'Farm 1' seems a natural choice. This choice turns out to be sufficient. In addition, when

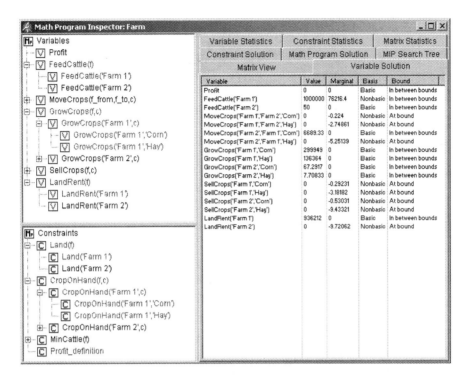

Figure 9.8: An identified substructure causing unboundedness

inspecting the bookmarked variables and constraints on the Matrix View tab, the red color of the objective function coefficient for the 'FeedCattle' variable for 'Farm 1' indicates a badly scaled value.

9.3.4 Analyzing an unrealistic solution

The example model is artificially turned into a model with an unrealistic solution by increasing the crop yield for corn on 'Farm 2' from 128 to 7168 (a mistake), and setting the minimum cattle requirement to zero. As a result, it will be unrealistically profitable to grow corn on 'Farm 2'.

Introducing an unrealistic solution

Once the changes from the previous paragraph have been applied, the solution of the model is shown in Figure 9.9. From the Variable Solution tab it can indeed be seen that the profit is unrealistically large, because a large amount of corn is grown on 'Farm 2', moved to 'Farm 1' and sold on 'Farm 1'. Other striking numbers are the large reduced cost values associated with the 'FeedCattle' variable on 'Farm 2' and the 'GrowCrops' variable for hay on 'Farm 2'.

Inspecting the unrealistic solution

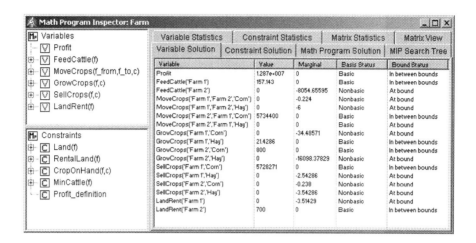

Figure 9.9: An unrealistic solution

Badly scaled matrix coefficients

When investigating an unrealistic solution, an easy first step is to look on the Matrix View tab to see whether there exist matrix coefficients with unrealistic values. For this purpose, first open the Matrix View tab in symbolic view. Blocks that are colored red indicate the existence of badly scaled values. By double clicking on such a block, you will zoom in to inspect the matrix coefficients at the individual level. In our example, the symbolic block associated with the 'GrowCrops' variable and the 'CropOnHand' constraint is the red block with the largest value. When you zoom in on this block, the data error can be quickly identified (see Figure 9.10).

Primal and dual contributions . . .

A second possible approach to look into the cause of an unrealistic solution is to focus on the individual terms of both the primal and dual constraints. In a primal constraint each term is the multiplication of a matrix coefficient by the value of the corresponding variable. In the Math Program Inspector such a term is referred to as the *primal contribution*. Similarly, in a dual constraint each term is the multiplication of a matrix coefficient by the value of the corresponding shadow price (i.e. the dual variable). In the Math Program Inspector such a term is referred to as the *dual contribution*.

. . . can be unrealistic

Whenever primal and/or dual contributions are large, they may indicate that either the corresponding coefficient or the corresponding variable value is unrealistic. You can discover such values by following an iterative process that switches between the Variable Solution tab and the Constraint Solution tab by using either the **Variable Statistics** or the **Constraint Statistics** command from the right-mouse popup menu.

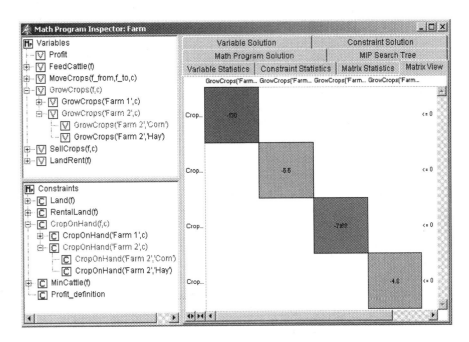

Figure 9.10: The Matrix View tab for an unrealistic solution

The following iterative procedure can be followed to resolve an unrealistic solution.

Procedure to resolve unrealistic solutions

- Sort the variable values retrieved through the Variable Solution tab.
- Select any unrealistic value or reduced cost, and use the right-mouse popup menu to switch to the Variable Statistics tab.
- Find a constraint with an unrealistic dual contribution.
- If no unrealistic dual contribution is present, select one of the constraints that is likely to reveal some information about the construction of the current variable (i.e. most probably a binding constraint).
- Use the right-mouse popup menu to open the Constraint Statistics tab for the selected constraint.
- Again, focus on unrealistic primal contributions and if these are not present, continue the investigation with one of the variables that plays an important role in determining the level value of the constraint.
- Repeat this iterative process until an unrealistic matrix coefficient has been found.

You may then correct the error and re-solve the model.

In the example, the 'Profit' definition constraint indicates that the profit is extremely high, mainly due to the amount of corn that is sold on 'Farm 1'. Only two constraints are using this variable, of which one is the 'Profit' definition itself. When inspecting the other constraint, the 'CropOnHand' balance, it shows that the corn that is sold on 'Farm 1' is transported from 'Farm 2' to 'Farm 1'.

Inspecting primal contributions

This provides us with a reason to look into the 'CropOnHand' balance for corn on 'Farm 2'. When inspecting the primal contributions for this constraint the data error becomes immediately clear (see Figure 9.11).

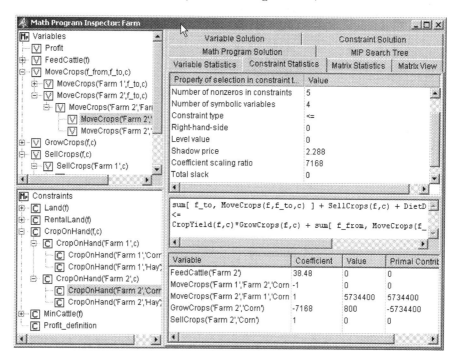

Figure 9.11: Inspecting primal contributions

Inspecting dual contributions

The same mistake can be found by starting from an unrealistic reduced cost. Based on the large reduced cost for the 'FeedCattle' variable on 'Farm 2', the dual contributions indicate that the unrealistic value is mainly caused by an unrealistic value of the shadow price associated with the 'Land' constraint on 'Farm 2'. While investigating this constraint you will notice that the shadow price is rather high, because the 'GrowCrops' variable for corn on 'Farm 2' is larger than expected. The dual contribution table for this variable shows a very large coefficient for the 'CropOnHand' constraint for corn on 'Farm 2', indicating the data error (see Figure 9.12).

Combining primal and dual investigation

The above two paragraphs illustrate the use of just primal contributions or just dual contributions. In practice you may very well want to switch focus during the investigation of the cause of an unrealistic solution. In general, the Math Program Inspector has been designed to give you the utmost flexibility throughout the analysis of both the input and output of a mathematical program.

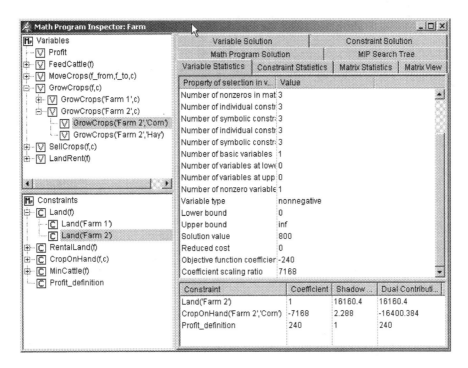

Figure 9.12: Inspecting dual contributions

Bibliography

[Ch91] J.W. Chinneck, *Locating minimal infeasible constraint sets in linear programs*, ORSA Journal on Computing, vol. 3, (1991), no. 1, 157–168.

[Mc98] B.A. McCarl, *A note on fixing misbehaving mathematical programs: Post-optimality procedures and GAMS-related software*, Journal of Agricultural & Applied Economics, vol. 30, (1998), no. 2, 403–414.

Part III

Creating an End-User Interface

Chapter 10

Pages and Page Objects

After you have created a model in AIMMS to represent and solve a particular problem, you may want to move on to the next step: creating a graphical end-user interface around the model. In this way, you and your end-users are freed from having to enter (or alter) the model data in ASCII or database tables. Instead, they can make the necessary modifications in a graphical environment that best suits the purposes of your model. Similarly, using the advanced graphical objects available in AIMMS (such as the Gantt chart and network flow object), you can present your model results in an intuitive manner, which will help your end-users interpret a solution quickly and easily.

Creating an end-user interface

This chapter gives you an overview of the possibilities that AIMMS offers you for creating a complete model-based end-user application. It describes pages, which are the basic medium in AIMMS for displaying model input and output in a graphical manner. In addition, the chapter illustrates how page objects (which provide a graphical display of one or more identifiers in your model) can be created and linked together.

This chapter

10.1 Introduction

A *page* is a window in which the data of an AIMMS model is presented in a graphical manner. Pages are the main component of an end-user interface for a model-based decision support application. An example of an end-user page is given in Figure 10.1. The page shown here provides a comprehensive graphical overview of the results of an optimization model by means of a *network flow object* in which flows which require attention are colored red. By clicking on a particular flow in the network object, additional information about that flow is shown in the tables on the left of the page.

What is a page?

Pages are fully designed by application developers for use by the end-users of an application. Thus, you, as a developer, can decide at what position in the interface particular model data should be presented to the end-user, and in which format. In addition, by automatically executing procedures when opening or closing a page or when modifying data, you can make sure that

Page design

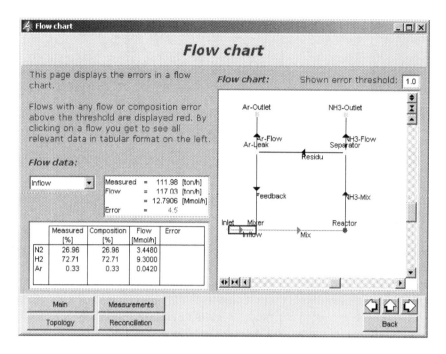

Figure 10.1: Example of a end-user page

all the necessary computations are performed before certain model results are displayed.

10.2   Creating pages

Creating pages

Creating an end-user page is as easy as adding a new node to the page tree in the **Page Manager** (see Chapter 12). Figure 10.2 illustrates the page tree associated with the example application used throughout this guide. As all the trees in the AIMMS modeling tools work alike, you can use any of the methods described in Section 4.3 to add a new page node to the page tree.

Copying pages

In addition to inserting a new empty page into the page tree, you can also copy an existing page or an entire subtree of pages, by either a simple cut, copy and paste or a drag-and-drop action in the tree (see Section 4.3). All copied pages will have the same content as their originals.

Page name

The node name of every page (as displayed in the page tree) is the unique name or description by which the page is identified in the system. When you add new pages to the tree, AIMMS will name these *Page 1, Page 2,* etc. You can change this name using the standard methods for changing names of tree nodes as described in Section 4.3.

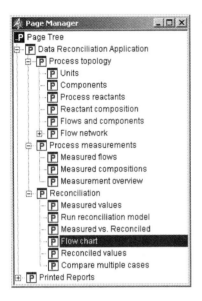

Figure 10.2: Example of a page tree

By default, the node name is the title that will be displayed in the frame of the page window when the page is opened. In the page **Properties** dialog box (see Section 11.2) you can, however, specify a different page title to be displayed, which can either be a constant string or a reference to a string parameter in the model. The latter is useful, for instance, if you intend to set up an end-user interface in multiple languages.

Page title

Its position in the page tree determines the navigational properties of the page. It will determine how any button with references to the next or previous page, or any navigation object or menu linked to the page, will behave. These navigational aspects of the **Page Manager** tool are discussed in more detail in Chapter 12.

Position in page tree

Every page that you add to the page tree, is also automatically added to the template tree in the **Template Manager**. By moving the page to a different position in the template tree, the page automatically inherits all the properties such as page size or background, and all objects specified on the template pages hierarchically above it. The **Template Manager** and the use of templates is explained in full detail in Chapter 12.

Using templates

10.3 Adding page objects

All visible components on a page are instances of the collection of *page objects* as offered by AIMMS. Such page objects are mostly used to visualize the

Page objects

input and output data of your model in various ways. They also include simple drawing objects, such as lines and circles, and buttons for navigation and execution.

Edit mode

Before you can add page objects to a page, the page must be in *edit mode*. When you open a page using the **Page Manager**, it is opened in *user mode* by default. When you want to open a page in edit mode from the **Page Manager**, you can do so using the right mouse pop-up menu. If a page is already opened in user mode, you can reopen it in edit mode using the 🖾 button on the page toolbar. When you open the page from the **Template Manager**, it is opened in edit mode by default.

Common data objects

AIMMS provides the most common graphical data objects such as

- row-oriented composite tables,
- 2-dimensional tables,
- pivot tables,
- graphs, and
- charts.

These objects can be used both for displaying and for modifying the data in your model. The data displayed in such objects are always directly linked to one or more identifiers in your model.

Adding an object

Placing a data object onto a page can be done without any programming. The following straightforward actions are required:

- select the type of the graphical object to be displayed,
- drag a rectangle onto the page of the intended size of the object, and
- choose the identifier in the model holding the data that you want to display.

Selecting the object type

You can select the object type that you want to add to the page from the **Object** menu. Alternatively, you can select any of the most common object types using the **Page Edit** toolbar, as depicted in Figure 10.3. If you move the cursor to one

Figure 10.3: The **Page Edit** toolbar

of the buttons of the toolbar, a tooltip will appear. After you have selected an object type, the page cursor will change to a cross allowing you to drag the rectangle in which the object will be contained. Figure 10.4 illustrates such a rectangle just prior to linking it to one or more AIMMS identifiers.

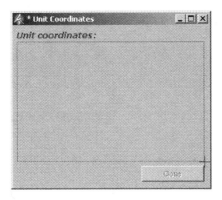

Figure 10.4: Drawing an object region

In order to let you drag object regions in an aligned manner, AIMMS allows you to associate a grid with a page, and align object regions to that grid automatically via the **View** menu. Alternatively, you may align objects later on, or make them the same size via the **Edit-Alignment** menu (see Section 11.1).

Object grid and alignment

After you have indicated the object region, you must select an identifier to be associated with that object. To support you in this task AIMMS provides an **Identifier Selection** dialog box as illustrated in Figure 10.5. You can select any single identifier from the list on the right.

Selecting an identifier ...

Additional help is offered for models with many identifiers. By selecting a subtree of the model tree on the left-hand side of the dialog box, you can narrow down the selection of identifiers on the right-hand side to those which are declared within the selected subtree. With the **Filter...** button you can narrow the selection down even more, by only displaying those identifier types that you are interested in.

... from a subselection

When your project contains one or more library projects, AIMMS only allows you to select identifiers that are part of the interface of a library on any page not included in such a library (see also Section 3.2). If the page is part of the page tree of a library, AIMMS allows you to select from *all* the identifiers declared in the library.

Selecting from a library

By restricting access from within pages outside of the library to the identifiers in the library interface only, AIMMS allows you to freely modify the internal implementation of your library. No other part of the application will be inflicted if you make changes to identifier declarations that are not included in the library interface.

Ensuring your freedom

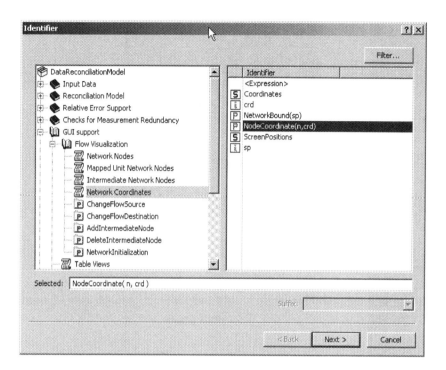

Figure 10.5: The **Identifier Selection** dialog box

Slices and
linking

In its simplest form, you can use the **Identifier Selection** dialog box to select an entire identifier of the appropriate dimension to fill a selected object. However, the **Identifier Selection** dialog box will also let you consider selecting *slices* of identifiers, or provide automatic *links* between objects. These advanced subjects will be discussed in detail in Section 10.4 below.

Object
properties

After you have selected the identifier(s) necessary to fill the page object with the appropriate model data, AIMMS will draw the object using default settings for properties such as fonts, colors and borders. Later on, you can change these properties (or even modify the defaults) via the **Properties** dialog box of the object (see also Section 11.2).

Example

If the object region displayed in Figure 10.4 is used to draw a table object, and the identifier selection dialog box in Figure 10.5 is used to select the identifier NodeCoordinate(n,crd), the table in Figure 10.6 results.

10.3.1   Displaying expressions in page objects

Displaying
expressions

In addition to indexed identifiers, AIMMS also allows you to display expressions in a page object. This is convenient, for instance, when you want to display some data which is not directly available in your model in the form of an

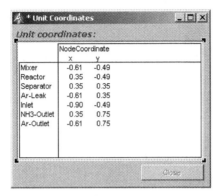

Figure 10.6: Example of a newly created table object

(indexed) identifier, but which can be easily computed through an expression referring to one or more identifiers in your model. In such a case, you do not have to create an additional defined parameter containing the expression that you want to display, but rather you can directly enter the expression in the **Identifier Selection** dialog box, as illustrated in Figure 10.7.

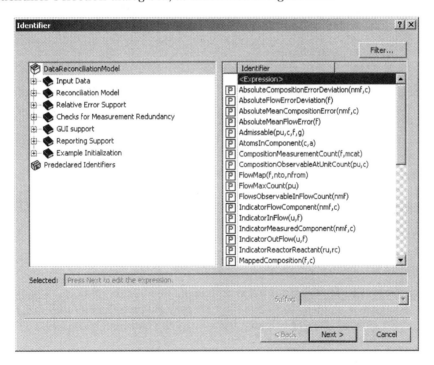

Figure 10.7: Selecting an expression in a page object

Entering an expression

When you have indicated that the page object should display an expression rather than an indexed identifier, AIMMS will display the **Expression Definition** dialog box illustrated in Figure 10.8. In this dialog box you must specify the exact definition of the expression you want to be displayed in the page object.

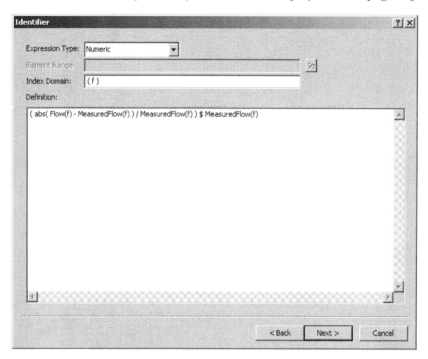

Figure 10.8: Entering an expression for a page object

Specifying the expression type

In the **Expression Type** field of the **Expression Definition** dialog box, you must select the type of the expression you entered. AIMMS only allows the display of

- numeric,
- element-valued, and
- string-valued.

expressions. AIMMS does not allow the display of set expressions. If the expression type is element-valued, you must also indicate the element range of the expression, i.e. the set in which the expression will hold its values.

Specifying the index domain

Finally, in the **Index Domain** field of the **Expression Definition** dialog box you must specify the index domain over which the expression is defined. Contrary to the INDEX DOMAIN attribute in a parameter declaration form, AIMMS only accepts a list of indices in this field, i.e. you cannot add a domain condition (see also Section 4.1 of the Language Reference). If you want to restrict the domain of the expression, you specify the domain condition as a $ condition within the expression definition (see also Section 6.1.9 of the Language Reference). This is illustrated in Figure 10.8, where MeasuredFlow(f) serves as a domain condition on the domain f.

10.3.2 Creating advanced page objects

In addition to common graphical data objects such as tables, bar charts and curves, AIMMS also supports a number of advanced graphical objects. These objects are designed for specialized, but widely-used, application areas. The most notable advanced objects available in AIMMS are:

Advanced data objects ...

- an interactive *Gantt chart* for time-phased scheduling and planning applications, and
- a *network flow object* for applications in which two-dimensional maps or flows play a central role.

Advanced data objects have the characteristic that multiple model identifiers are required to represent the visual result. For instance, in the network flow object you need a set identifier to denote the set of nodes to be displayed and their coordinates in the network, as well as a parameter to indicate the flow values between these nodes. Figure 10.9 illustrates the selection dialog box of a network flow object. To enter the appropriate identifiers for each

... are based on multiple identifiers

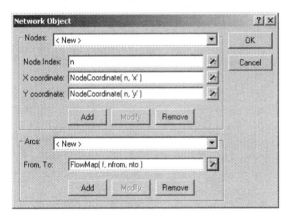

Figure 10.9: Identifier selection for the network flow object

required component, you can open the common **Identifier Selection** dialog box described above by pressing the wizard button at the right of each individual component.

Object help   In this User's Guide you will only find the basic mechanisms for adding or modifying pages and page objects. Full details of all object types, and their properties and settings, are described in the on-line help facility which is always available when you are running AIMMS.

Non-data objects   In addition to data-related objects, AIMMS also supports various other types of objects such as:

- drawing objects (such as line, circle, rectangle, picture and text objects), and
- buttons to initiate model execution and page navigation.

Drawing objects and buttons are positioned on a page in exactly the same manner as the data objects described above, except that a link to one or more AIMMS identifiers is not required.

10.4 Selecting identifier slices and linking objects

Advanced identifier selection   After you have selected an indexed identifier (or expression) in the **Identifier Selection** dialog box, a second dialog box appears, as illustrated in Figure 10.10. In this dialog box, you have several options to refine your choice,

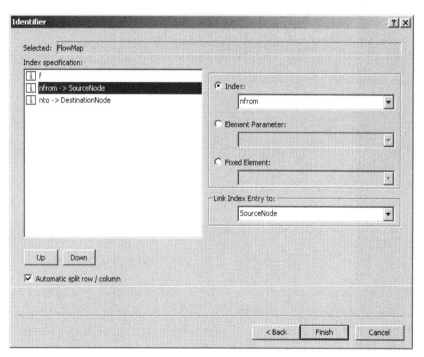

Figure 10.10: Advanced **Identifier Selection** options

each of which will be described in this section.

By default, AIMMS assumes that you want to associate the full identifier with the object in hand. However, with the dialog box of Figure 10.10 AIMMS allows you to modify several domain-related issues before displaying the identifier. More specifically, for every individual dimension in the index domain of the identifier, you can:

Slicing and subset restriction

- restrict that dimension to those elements that are included in a particular subset associated with the domain set by using a subset index,
- select a slice of the identifier by fixing that dimension to the value of a particular scalar element-valued parameter that assumes its values into the corresponding domain set, or
- select a slice of the identifier by fixing that dimension to a specific element in the corresponding domain set.

In the dialog box of Figure 10.10 AIMMS lets you select specific elements, element parameters or subset indices on the right-hand side of the dialog box to restrict the dimension that is selected on the left-hand side.

By fixing a particular dimension to an element parameter or a set element, the total number of dimensions of the displayed data is reduced by one. Thus, by fixing one dimension of a two-dimensional parameter, only a one-dimensional table will be displayed. The number of dimensions is not reduced when the display is restricted to elements in a subset. In this case, however, the object will display less data.

Dimension reduction

For a table object, the **Identifier Selection** dialog box also lets you determine the order of the dimensions and a split of the dimensions. This allows you to specify which dimensions are shown rowwise and which columnwise, and in which order. If you do not insert a split manually, AIMMS will determine a default split strategy.

Index order and table split

Finally, the identifier selection options offer you the possibility of establishing a link between a particular dimension of the selected identifier and a (scalar) element parameter that assumes its values into the corresponding domain set. As an example, consider the dialog box of Figure 10.10. In it, the dimension nfrom of the identifier FlowMap(f,nfrom,nto) is linked to the element parameter SourceNode, and the dimension nto to the element parameter DestinationNode.

Index linking

In the **Properties** dialog boxes of a linked object, AIMMS displays the link using a "->" arrow. Thus, the parameter FlowMap from the example above, will be displayed as

Link notation

```
FlowMap( f, nfrom -> SourceNode, nto -> DestinationNode )
```

This special link notation is only valid in the graphical interface, and cannot be used anywhere else in the formulation of your model.

Effect of index linking

When the identifier FlowMap(f,nfrom,nto) is displayed in, for instance, a table object, AIMMS will, as a result of the specified index links, automatically assign the values of nfrom and nto associated with the currently selected table entry to the element parameters SourceNode and DestinationNode, respectively.

Use of index linking

Index linking is a very powerful AIMMS feature that allows you to effectively implement several attractive features in an end-user interface without any programming effort on your part. Some representative uses of index linking are discussed below.

- You can use index links involving one or more element parameters in a particular page object as a way of triggering AIMMS to automatically update one or more other page objects that contain identifier slices fixed to these element parameters. These updates will occur as soon as a user clicks somewhere in the particular page object in which the indices were linked. An illustrative example of such automatic linkage of page objects is shown below.
- You can use index linking to keep track of the current user selection in an object when executing a procedure within your model. This allows you to do some additional data processing, or perform some necessary error checks for just that tuple in a multidimensional identifier, whose value has most recently been modified by the end-user of your application.

Example

Consider the page shown in Figure 10.11. The tables and lists in the left part

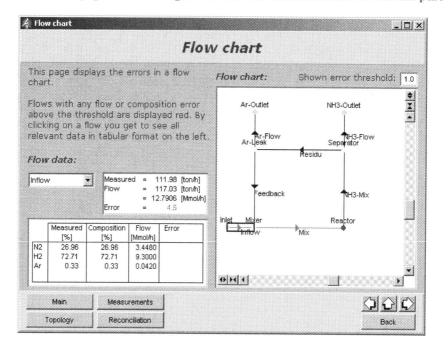

Figure 10.11: Example of index linking

of the page display detailed information regarding the currently selected flow in the network flow object shown in the right part of the page. This was accomplished as follows. The index f representing the flows in the network flow object on the right was linked to a single element parameter FlowEl in the set Flows. The tables and lists on the left of the screen contain identifier slices fixed to the element parameter FlowEl. Take, for instance, the values in the column named Measured in the table object on the lower left part of the screen. This column corresponds to the one-dimensional identifier slice MappedMeasuredComposition(c,FlowEl). As a result of the link, the column Measured automatically displays detailed information for the flow selected by the end- user in the flow chart on the right.

Chapter 11

Page and Page Object Properties

This chapter

After you have created a page with one or more data objects on it, AIMMS allows you to modify the display properties of these objects. This chapter illustrates the available tools for placing and ordering page objects, and how to modify properties of both pages and page objects. It also provides a brief description of the available properties.

11.1 Selecting and rearranging page objects

Selecting an object

Before you can modify the properties of a page object, you must select the object. This can be accomplished as follows:

- make sure that the page is opened in edit mode (see Section 10.3),
- press the **Select Object** button ![button] on the page toolbar, if it is not already pressed, and
- click on the page object.

The selected object(s) on a page are marked with a small dark square on each of its corners. This is illustrated in Figure 11.1.

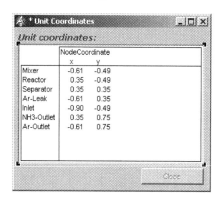

Figure 11.1: A selected page object

When a page depends on one or more templates (see also Section 12.2), AIMMS will only let you select those objects that were placed on the page itself, and not those which are contained in any of its templates. Template objects can only be edited in the template page on which they are defined.

No template objects

When two or more objects are overlapping, clicking on the overlapping region will result in any one of the overlapping objects being selected. By holding the **Shift** key down during clicking, AIMMS will cycle through all the overlapping objects, allowing you to select the object of your choice. Alternatively, you can press the **Tab** key repeatedly to browse through all selectable objects on the page.

Selecting overlapping objects

In addition to selecting a single page object, AIMMS also allows you to select multiple objects. You can do this by dragging a *select region* on the page, after which AIMMS will mark all objects contained in that region as selected. Alternatively, you can add or remove objects to form a selection by clicking on the objects while holding down the **Shift** key.

Selecting multiple objects

With the **Edit-Alignment** menu of a page in edit mode, you can correct the placement and sizes of all page objects that are currently selected. The **Alignment** menu lets you perform actions such as:

Object alignment

- give all selected objects the same height or width, i.e. the height or width of the largest object,
- align all selected objects with the top, bottom, left or rightmost selected object,
- center the selected objects horizontally or vertically, and
- spread all selected objects equally between the top and bottommost objects or between the left and rightmost objects.

An alternative method of alignment is to define a grid on the page (see Section 10.3), and align the borders of all objects with the grid.

With the **Drawing Order** item of the **Edit** menu, you can alter the order in which overlapping objects are drawn. When applied to a selected object, you can specify that the object at hand must be drawn as either the top or bottommost object. Modifying the drawing order only makes sense for drawing objects such as the text, rectangle, line, circle and picture objects.

Drawing order

When there is a natural order in which an end-user has to enter data on a particular page, you can use the **Tab Order** item from the **Edit** menu, to specify this order. The **Tab Order** menu opens a dialog box as illustrated in Figure 11.2. In this dialog box all page objects are displayed in a list which determines the (cyclic) order in which AIMMS will select the next object for editing when the user leaves another object on the page through the **Tab** or **Enter** keys.

Specifying the tab order

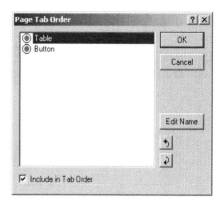

Figure 11.2: The **Tab Order** dialog box

Tabular objects

In tabular objects, the **Tab** and **Enter** keys can also be used to move to the next table entry to the right or below, respectively. In such cases, AIMMS will only go to the next object in the tab order, if further movement to the right or below within the object is no longer possible.

Disabling tab order

In addition to modifying the tab order, you can also use dialog box of Figure 11.2 to select the page objects that should not be included in the tab order. Alternatively, you can remove a page object from the tab order in the **Properties** dialog box of that object as explained in the next section. Objects excluded from the tab order are not accessible on the page by pressing the **Tab** or **Enter** keys, but can still be selected using the mouse.

11.2   Modifying page and object properties

Object properties

In addition to modifying the display properties of groups of objects on a page, AIMMS also allows you to modify the visual appearance of a page itself and of all of its individual page objects. When the page is in edit mode, you can open the **Properties** dialog box of either a page or a page object by simply double clicking on it, or by selecting **Properties** from the right-mouse pop-up menu. This will display a dialog box as illustrated in Figure 11.3. The dialog box contains tabs for all visual aspects that are relevant to that object, and initially displays the current settings of these visual aspects.

Properties of multiple objects

You can also modify properties of multiple objects at the same time by first selecting a group of objects and then selecting the **Edit-Properties** menu, or selecting **Properties** from the right-mouse pop-up menu. This will invoke a **Properties** dialog box containing only those tabs that are common to all the selected objects. AIMMS will not display an initial value for the corresponding properties, as each property may hold different initial values for the various

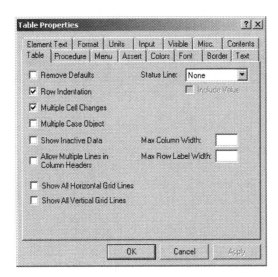

Figure 11.3: The **Properties** dialog box

objects. Only the properties that you change are applied to the selected objects.

Through the tabs in the **Properties** dialog box, AIMMS lets you modify the various properties of pages and page objects. The following paragraphs provide a brief overview of the modifiable properties. A full explanation of the various properties of all the available objects can be found in the help file accompanying the AIMMS system.

Property types

With the **Contents** tab you can add or remove identifiers from the list of identifiers that are displayed in the object. With this tab you can specify, for instance, that a table is to display the values of two or more identifiers. To modify the contents, AIMMS will open the common **Identifier Selection** dialog box as explained in Section 10.3.

The Contents tab

Before you can make changes to the **Contents** tab, AIMMS requires that you apply any changes you have made to the other object properties before entering the **Contents** tab. You can apply these changes using the **Apply** button. Similarly, after you have made changes to the **Contents** tab, AIMMS requires that you apply these changes before you can go on to modify other object properties.

Applying changes

With the **Procedure** tab you can specify the particular procedures that must be executed upon user inputs such as a data change or selecting a particular value in a data object. The use of procedures linked to data objects is mostly to perform error checks or update other identifiers based on a single data change.

The Procedure tab

The Action tabs   With the **Action** tab, the counterpart of the **Procedure** tab for pages, buttons and navigational controls, you can specify the particular actions that must be executed upon opening a page, pressing a button, or making a selection in a navigational control. Such actions typically can be a sequence of running a procedure within the model, executing predefined AIMMS menu actions, or checking assertions.

The Menu tab   The **Menu** tab lets you specify which menu bar, toolbar, and right-mouse pop-up menu should be active on top of either a page or an object on a page. The menus themselves, as well as the actions linked to the menus, can be created in the **Menu Builder** tool. The **Menu Builder** tool is explained in full detail in Chapter 12.

Double-click   An action type that is used quite frequently, is the double-click action. You
actions   can specify a double-click action either in the **Action** tab, or through the **Menu** tab. The following rules apply.

■ If a **Double-Click** procedure is specified on the **Action** tab, AIMMS will execute that procedure.

■ If no **Double-Click** procedure has been specified, but a pop-up menu associated with the page object has a default item, AIMMS will execute the default menu item.

■ If neither of the above apply, and the object is a table displaying a set, the double-click action will toggle set membership of the set element which currently has the focus.

■ In all other cases, double-clicking will be ignored.

The Assert tab   Through the **Assert** tab you can indicate which assertions already declared in your model are to be checked upon end-user data changes to a particular identifier in a data object. AIMMS can perform the assertion immediately upon every data change, or delay the verification until the end-user presses a button on the page. Once an immediate assertion fails, the assertion text will be displayed to the user and the original value will be restored.

The Colors tab   With the **Colors** tab you can not only specify the colors that are to be used for the foreground and background of a page or page object, but also the color for the user-selected values in a page object. In addition, you can specify a model-defined (indexed) color parameter to define the foreground color that will be used for each identifier in a data object. With such a parameter you can, for instance, color *individual* values of an object depending on a certain threshold. The necessary computations for this individual coloring need to be made inside the model underlying the end-user interface. You will find more details about assigning color parameters in Section 11.4.

The **Font** tab lets you define the font that is to be used for a particular object. *The Font tab*
You can choose the font from a list of user-defined font descriptions as illustrated in Figure 11.4. To add a new font name to the list, you should press the

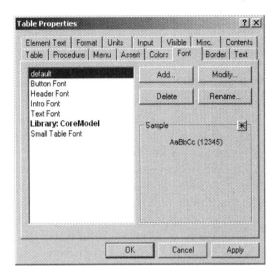

Figure 11.4: The **Font** tab of a **Properties** dialog box

Add button. This will open the standard Windows font selection dialog box,
allowing you to define a new AIMMS font based on the list of available Windows fonts, font styles and sizes. Once you have made a selection, you will be
requested to provide a description for the newly selected font.

It is strongly recommended that you choose functional names for AIMMS fonts   *Choose*
(i.e. describing their intended use) instead of merely describing the choices you   *functional font*
made. For instance, naming a new font "Button font" instead of "Arial Regular,   *names*
8 pt" will help tremendously in preventing mistakes when selecting a font for
a button.

AIMMS also allows you to store fonts within a library project. The list of fonts   *Fonts in library*
shown in Figure 11.4 displays a single font *Small Table Font* associated with   *projects*
the library *CoreModel*. You can manage the list of fonts associated with a
library by pressing the buttons on the right-hand side of the dialog box, while
the selection in the listbox on the left-hand side is in the area associated with
the library.

If you have defined fonts within a library project, you should ideally only use   *Use only in*
these fonts in pages that are also part of the library project. If you use the   *library pages*
fonts in pages outside of the library, such pages may fail to display properly
after you have removed the library project from the AIMMS project.

Font names must be unique

AIMMS requires that all font names be unique across the main project and all library projects that are included in the main project. If you include an existing library project, which contains a font name that is already present in the AIMMS project, AIMMS assumes that both fonts are the same and will ignore the second font definition.

The Border tab

With the **Border** tab you can stipulate the border settings for any particular data object on a page. A border can consist of merely a surrounding line, or provide an in- or out-of shadow effect.

The Text tab

With the **Text** tab you can specify for each identifier a single line of text that is displayed in a page object. With this line of text you can, for instance, provide descriptions for the data in a table containing one or more identifiers. In addition, the **Text** tab will let you define the element description for the (optional) status line associated with the object. The status line will display the currently selected value along with its element description. If the element description contains references to the indices over which the identifier at hand is defined, these references will be expanded to the currently selected element names.

The Element Text tab

By default, any set element in a data object will be displayed by its name in the model. If you want to display an alternative text for a set element, you can use the **Element Text** tab to specify a string parameter holding these alternative element descriptions. You can use this feature, for instance, to display set elements with their long description in the end-user interface, whereas the model itself, and perhaps paper reports, work with short element names.

The Format tab

The **Format** tab defines the numerical format in which the data of a particular identifier is displayed. This format can be specified on the spot, or can use a named format already predefined by you as the application developer. The display format specifies not only such properties as the width of a number field and its number of decimal places, but also their relative alignment, the use of a 1000-separator for large numbers, and the display of default values.

The Units tab

The AIMMS modeling language offers advanced support for defining units of measurement for each identifier in the model. In particular, AIMMS supports unit conventions which let you define a coherent set of units (e.g. Imperial or metric units) in a single declaration. In the end-user interface you can indicate in the **Units** tab whether you want units to be displayed for every identifier or for every individual value contained in a particular data object. The displayed units are the units defined for the identifier at hand, unless the end-user has selected a current unit convention with alternative units. Figure 11.5 illustrates an end-user page in which identifier values are displayed along with their associated units of measurement.

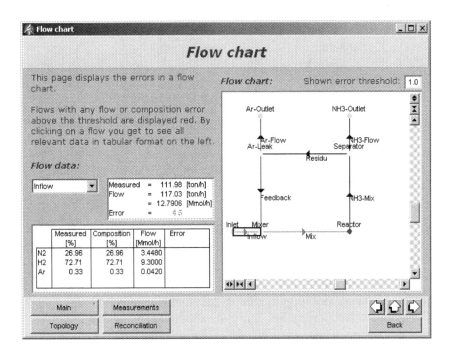

Figure 11.5: Use of units in a data object

With the **Input** tab you can specify the read-only properties of every identifier *The Input tab* in a page object separately. The decision as to whether numbers are read-only can depend on (indexed) identifiers in your model. Thus, you can arrange it so that particular numbers in, for example, a table can be edited by the end-user, while other numbers associated with that same identifier are considered as read-only. In addition to the properties specified on this tab, the overall read-only behavior of identifiers is also influenced by the contents of the predefined identifier CurrentInputs (see Section 18.1).

You can use the **Visible** tab to hide a particular page object in its entirety *The Visible tab* from a page. Whether or not a page object is visible may depend on a scalar identifier (slice) in your model. The ability to hide page objects comes in handy when, for instance,

- you want to hide a page object because a particular end-user has no right to modify its data, or
- a page contains two exactly overlapping page objects—e.g. one holding relative numbers, the other holding absolute numbers—and you want to display just the one based on the user's choice.

With the **Misc.** tab you can specify various miscellaneous settings such as *The Misc. tab*

- whether a page object must be included in the page tab order to specify a natural navigation order on the page (see also Section 11.1),

- whether an object is actually printed or skipped during printing (only relevant for print pages, see also Chapter 14),
- which end-user help topic should be displayed for the page or page object at hand, or
- a tag name, which is used when you want to refer to the object from within the model (see Section 18.4.1).

Help file

Before adding end-user help to a particular page, page object, end-user menu or toolbar, you must add a help file to your project directory, and specify its name through the **Options** dialog box (see Section 21.1). All the available end-user help associated with your project must be contained in the specified project help file.

Help file formats

AIMMS supports several help file formats, allowing you to create a help file for your project using the tools you are most familiar with. They are:

- standard Windows help files (with the .hlp extension),
- compiled HTML help files (with the .chm extension), and
- PDF files (with the .pdf extension), which require that Acrobat Reader version 4.0 or higher is installed on your machine.

An executable Acrobat Reader installation can be downloaded from the Adobe website www.adobe.com.

Creating help files

To create a help file in any of the supported formats you will need an appropriate tool such as RoboHelp, Help & Manual or DocToHelp to create either a Windows or compiled HTML help file, or Adobe Acrobat to create a PDF file. To jump to a marked position inside the help file when providing help for a page, a page object, a menu or a button on a toolbar you should add:

- (so called) *K-keywords* to an ordinary Windows help file,
- *keywords* to a compiled HTML help file, or
- *named destinations* added to a PDF file.

All of the destinations that you added to the help in this way can serve as the **Help Topic** inside the **Misc.** tab of a page or page object.

Object-dependent properties

In addition to the tabs described above, which are common to most objects, the **Properties** dialog box also has a number of tabs where you can change properties that are very specific for a particular type of object. Through such object-dependent options you can specify, for instance, whether a table should display default values, what should be displayed along the axes in a graph or chart, or how the arcs and nodes in a network flow object should be drawn. The contents of these object-specific tabs are explained in full detail in the online AIMMS help file.

11.3 Using pages as dialog boxes

By default all end-user pages behave as normal windows, i.e. whenever you
have multiple windows open, you can freely switch from window to window
simply by clicking in the window that should become active. Sometimes, however, your application may contain sequential actions which require the user
to make a certain choice or data change before moving on to the next action. In
this case the page should behave as a dialog box instead of a normal window.
While a dialog box is displayed on the screen, it is impossible to access other
windows in the application without closing the dialog box first for example
with an **OK** or **Cancel** button. By using dialog boxes you can force an end-user
to follow a strict sequence of operations.

*Use of dialog
boxes*

In AIMMS you can define that a page should behave like a dialog box by using
the page **Properties** dialog box as illustrated in Figure 11.6. If such a dialog

Dialog pages

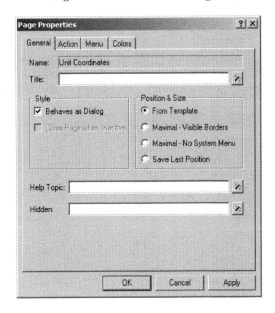

Figure 11.6: Creating a dialog page

page is opened using either a button, a menu, a navigation object or from
within the model through a call to the PageOpen procedure, it will behave like
a dialog box. *If, on the other hand, the dialog page is opened from within
either the **Page Manager** or the **Template Manager**, the page will behave as
an ordinary window.* This offers you the possibility of editing the contents and
layout of the page.

Blocking
execution

When a dialog page is called from within a procedure using PageOpen, the execution of the calling procedure will only continue after the dialog page has been closed by the end-user. In this way, any data supplied by the end-user in the dialog page will always be available during the remaining execution of the calling procedure.

Dialog box
result

Note that dialog pages do not offer built-in support to determine whether an end-user has finished the dialog box for example by pressing the **OK** or **Cancel** button. However, such control can easily be modeled in the AIMMS language itself. Perhaps the most straightforward manner to accomplish this is by introducing

- a set DialogActions containing two elements 'OK' and 'Cancel',
- an associated global element parameter CurrentDialogAction, and
- procedures such as ButtonOK and ButtonCancel which set CurrentDialog-Action equal to 'OK' or 'Cancel', respectively.

Linking to
dialog box
buttons

To obtain the result of a dialog page, you can simply add the execution of the procedures ButtonOK or ButtonCancel to the list of actions associated with the **OK** and **Cancel** buttons, respectively. In addition, you should link the functionality of the close icon for the dialog page to that of the **Cancel** button as illustrated in Figure 11.7.

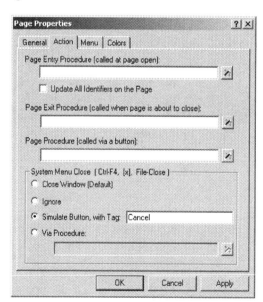

Figure 11.7: Linking dialog close to **Cancel**

To obtain the end-user choice in the dialog page after the return of the PageOpen procedure, you can simply check for the value of the element parameter CurrentDialogAction, as illustrated in the following code excerpt.

Obtaining the result

```
! Open the dialog page and stop processing when the user
! has pressed the 'Cancel' button.

OpenPage( "Supply input data" );
return 0 when CurrentDialogAction = 'Cancel';

! Otherwise perform further data processing based on the supplied input data.
```

You may want to create a customized dialog page *template* (see also Section 12.2) to capture the end-user choices as described above. Based on such a dialog page template, you can quickly create as many dialog pages as necessary, all behaving in a similar fashion when opened in a procedure of your model.

Create a dialog page template

11.4 Defining user colors

As already explained in the previous section, AIMMS allows you to define the color of particular objects in a graphical end-user interface from within the execution of your model. In this section you will see how you can define *user colors* which can be used within the model, and how you can use them to provide model-computed coloring of page objects.

User colors

To define user colors that persist across multiple sessions of a project, you should open the **User Colors** dialog box as illustrated in Figure 11.8 from the **Tools** menu. By pressing the **Add** or **Change Color** button, AIMMS will display the standard Windows color selection dialog box, which you can use to create a new user color or modify an existing user color. After you have selected a color, AIMMS will request a name for the newly defined color for further usage within the model.

Defining persistent user colors

As with font names, you may prefer to choose functional color names rather than names describing user colors. For instance, colors named "Full tank color", "Partially filled color" and "Empty tank color" may be a much better choice, from a maintenance point-of-view, than such simple names as "Red", "Blue" and "Green". In addition, choosing descriptive names may make the intention of any assignment to, or definition of, color parameters in your model much clearer.

Functional color names

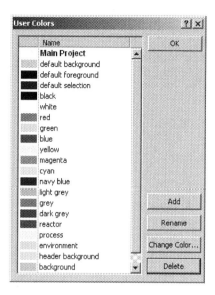

Figure 11.8: The **User Colors** dialog box

User colors in library projects

Similar as with fonts, a library project can also contain its own set of user colors. The list of colors shown in Figure 11.8 displays the user colors defined within the main project. For each library included in the project the listbox contains a separate area displaying the user colors that are associated with that library. You can manage the list of user colors associated with a library by pressing the buttons on the right-hand side of the dialog box, while the selection in the listbox on the left-hand side is in the area associated with the library.

Use only in library pages

If you have defined user colors within a library project, you should ideally only use these user colors in pages that are also part of the library project. If you use the user colors in pages outside of the library, such pages may fail to display properly after you have removed the library project from the AIMMS project.

Color names must be unique

AIMMS requires that all user color names be unique across the main project and all library projects that are included in the main project. If you include an existing library project, which contains a user color name that is already present in the AIMMS project, AIMMS assumes that both user colors are the same and will ignore the second color definition.

Runtime user colors

Persistent user colors cannot be modified or deleted programmatically. However, you can add runtime colors (which only exist for the duration of a project session) programmatically from within your model using the function User-ColorAdd. In the **User Colors** dialog box, runtime colors are shown under the header **Runtime colors**. You can modify or delete such runtime colors using the functions UserColorModify and UserColorDelete. These functions are discussed in full detail in Section 18.2.1.

All persistent and non-persistent user colors are available in your model as elements of the predefined set AllColors. To work with colors in your model you can simply define scalar and/or indexed element parameters into the set AllColors. Through simple assignments or definitions to such parameters you can influence the coloring of identifiers or individual identifier values on an end-user page.

The set AllColors

Consider a set of Flows in a network with index f. If a mathematical program minimizes the errors in computed flows in respect to a set of measured flow values, then the following simple assignment to a color parameter FlowColor(f) marks all flows for which the error exceeds a certain threshold with an appropriate color.

Example

```
FlowColor(f) := if ( FlowError(f) >= ErrorThreshold ) then
    'Red' else 'Black' endif;
```

With the above assignment, any graphical display of Flows can be colored individually according to the above assignment by specifying that the color of the individual numbers or flows in the **Colors** dialog box of the object be given by the value of the color parameter FlowColor(f). Figure 11.5 (on page 147) illustrates an example of an end-user page where the flows in the network flow object, as well as the individual entries in the tables and lists, are colored individually with respect to the parameter FlowColor(f) (the colors are only visible in the electronic version of this book).

Use in interface

Chapter 12

Page Management Tools

This chapter   When your decision support system grows larger, with possibly several people developing it, its maintainability aspects become of the utmost importance. To support you and your co-workers in this task, AIMMS offers several advanced tools. As discussed in Chapters 4 and 7, the **Model Explorer** combined with the **Identifier Selector** and **View Manager**, provide you with various useful views of the model's source code. In this chapter, the specialized AIMMS tools that will help you set up an advanced end-user interface in an easily maintainable manner will be introduced.

12.1 The Page Manager

Page navigation   In large decision support systems with many pages, navigating your end-users in a consistent manner through all the end-user screens is an important part of setting up your application. One can think of several organizational structures for all the available end-user screens in your application that would help your end-users maintain a good overview of their position (see also Chapter 15 for some background on designing end-user interfaces). To help you set up, and modify, a clear navigational organization quickly and easily, AIMMS provides a tool called the **Page Manager**.

The Page Manager   With the **Page Manager** you can organize all the existing pages of an AIMMS application in a tree-like fashion, as illustrated in Figure 12.1. The tree in the **Page Manager** that holds all the pages of the main AIMMS project is called the main *page tree*. Relative to a particular page in this page tree, the positions of the other pages define common page relationships such as *parent* page, *child* page, *next* page or *previous* page.

Library page trees   In addition to the main page tree, each library project included in the main project can have its own tree of pages as illustrated in Figure 12.1. Section 12.1.1 discusses the facilities available in AIMMS that allow you to combine the page structures in all trees to construct a single navigational structure for the entire application.

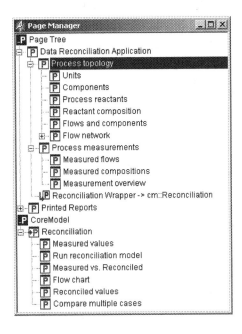

Figure 12.1: The **Page Manager**

The page relationships defined by the page tree can be used in several navigational interface components that can be added to a page or end-user menu. These components include

Navigational controls

- *navigation objects,*
- *navigation menus,*
- *button actions,* and
- *tabbed pages.*

These allow you to add dynamic navigation to the parent, child, next or previous pages with respect to the position of either

- the current page, or
- a fixed page in the page tree.

Section 12.1.2 explains in detail how to set up such automatic navigation aids.

The strength of the **Page Manager** tool lies in the fact that it allows you to quickly add pages to the page tree, delete pages from it, or modify the order of navigation without the need to make modifications to hard-coded page links on the pages themselves. Thus, when a model extension requires a new section of pages, you only need to construct these pages, and store them at the appropriate position in the page tree. With the appropriate navigational interface components added to the parent page, the new page section will be available to the end-user immediately without any modification of existing pages.

Aimed at ease of maintenance

12.1.1   Pages in library projects

Pages in library projects

AIMMS allows you to develop a separate page tree for every library project included in an AIMMS application. This is an important feature of library projects because

- it allows a developer to implement a fully functional end-user interface associated with a specific sub-project of the overall application completely independently of the main project, and
- pages defined inside a library project can refer to all the identifiers declared in that library, whereas pages defined in the main project (or in any other library) can only refer to the public identifiers in the interface of that library (see Section 3.2).

Duplicate names may occur

While AIMMS requires that the names of pages within a single (library) project be unique, page names need not be unique across library projects. To ensure global uniqueness of page names in the overall application, AIMMS internally prefixes the names of all the pages contained within a library with its associated library prefix (see Section 3.1). When you want to open an end-user page programmatically, for instance through the PageOpen function, you need to provide the full page name including its library prefix. Without a library prefix, AIMMS will only search for the page in the main page tree.

Separate page trees

The page trees associated with the main project and with all of its library projects are initially completely separate. That is, any navigational control (see Section 12.1.2) that refers to parent, child, next or previous pages can never navigate to a page that is not part of the tree in which the originating page was included.

All pages are accessible

Other than for the identifier declarations in a libray, AIMMS puts no restriction on which pages in the library can and cannot be shown from within the main application, or from within other libraries. Stated differently, the page tree of a library does not currently have a public interface.

Creating an application GUI

If an AIMMS project is composed of multiple libraries, then each of these libraries contains its own separate page tree, which may be combined to form the total end-user interface of the overall application. The navigational controls offered by AIMMS, however, can only reach the pages in the same tree in which an originating page is included.

Without further measures, pages from different libraries would, therefore, only be accessible through a unidirectional direct link, which is very undesirable from an end-user perspective. After following such a link moving to a parent, next or previous page may give completely unexpected results, and getting back to the originating page may be virtually impossible. For both developers and end-users a situation in which all relevant pages can be reached from within a single navigational structure is much more convenient.

Jumping to library pages

To address this problem, AIMMS offers a linkage mechanism called *page tree references*. Through a page tree reference, you can *virtually* move a subtree of pages from a library project to another location in either the main project or any other library project included in the AIMMS application. While physically the pages remain at their original location, the navigational controls in AIMMS will act as if the tree of pages has been moved to the destination location of the page tree reference. At the original location AIMMS' navigational controls will completely disregard the linked page tree.

Page tree references

You can create a page tree reference by inserting a page tree reference node at the destination location through the **Edit-New-Page Tree Reference** menu. In figure 12.1 the *Reconciliation Wrapper* node illustrates an example of a page tree reference node. It is linked to the tree of pages starting at the *Reconciliation* page in the *CoreModel* library. Note that AIMMS uses a special overlay of the page icon to visualize that a page is linked to a page tree reference node, and hence, at its original location, is not reachable anymore through AIMMS' navigational controls.

Creating a page tree reference

To create a link between a page tree reference node and a subtree of pages anywhere else in your application you have to select both the page tree reference node and the node that is at the root of the subtree that you want to link to, and select the **Edit-Page Tree Reference-Link** menu. You can unlink an existing link through the **Edit-Page Tree Reference-Unlink** menu.

Linking a page tree reference

12.1.2 Navigational interface components

The page tree can be used to directly control the navigational structure within an AIMMS-based end-user application. This can be accomplished either by special button actions or through the navigation object and menus. As an example, Figure 12.2 illustrates the *Process Topology* page contained in the page tree of Figure 12.1.In the lower right corner, the page contains three navigational buttons that are linked, from left to right, to the previous, parent and next page. Above this, the page contains a navigation object which, in this instance, automatically displays a list of buttons that corresponds exactly to the set of direct child nodes of the *Process Topology* page in the page tree.

Navigational control

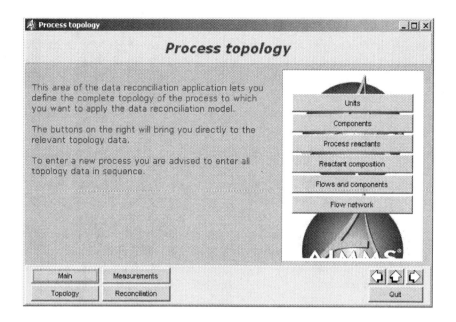

Figure 12.2: Page containing navigation buttons and a navigation object

Button actions   To add a page tree-based navigational control to a button, you only need to add a **Goto Page** action to the **Actions** tab in the button **Properties** dialog box, as illustrated in Figure 12.3. You can request AIMMS to open the previous,

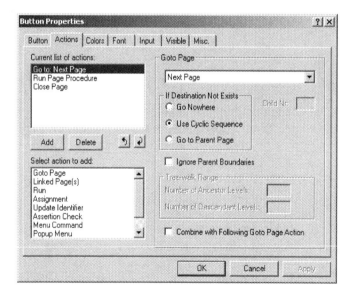

Figure 12.3: Adding navigational control to a button

next, parent or (first) child page relative to the position of the current page in the page tree. If you want the current page to be closed after opening the new page, you should additionally insert a **Close Page** action as in Figure 12.3.

When there is no longer a next or previous page to open in a particular branch *Cycling*
of a page tree, AIMMS will cycle to the first or last page within that branch,
respectively. You can further modify the result of a previous or next page
action by placing special *separator* nodes into the page tree, given that AIMMS
will never jump past a separator node. You will find the full details of separator
nodes in the online help on the **Page Manager**.

The second way to include a navigational control in an end-user page is by *Navigation*
means of a custom *navigation* object. A navigation object can display a subtree *object*
of the entire page tree in several formats, such as:

- a list of buttons containing the page titles (as in Figure 12.2),
- a list of buttons accompanied by the page titles,
- a list of clickable or non-clickable page titles without buttons, or
- a tree display similar to the page tree itself.

After adding a navigation object to a page, you must specify the subtree to *Object*
be displayed through the **Properties** dialog box as displayed in Figure 12.4. *properties*
What is displayed in the navigation object is completely determined by the

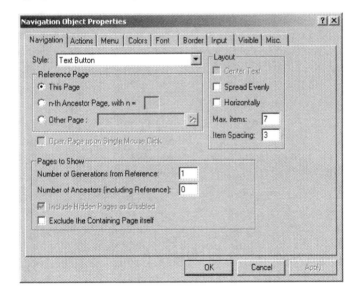

Figure 12.4: Navigation object **Properties** dialog box

reference page, together with the number of ancestor (parent) and child gen-
erations specified in this dialog box.

Display only   If you set a navigation object to read-only using the **Input** tab of the **Properties** dialog box, then you can use the navigation object for display- only purposes. Thus, you can use it to display the current page title as a page header, or the title of one or more parent pages in the header or footer area of the page. The "*Process Topology*" page header of the end-user page displayed in Figure 12.2 is an example of a display-only navigation object.

Navigation menus   Finally, you can add special navigation (sub)menus to your application in which the menu items and submenus represent a subtree structure of the page tree. Figure 12.5 illustrates an example of a navigation menu linked to the page tree displayed in Figure 12.1.

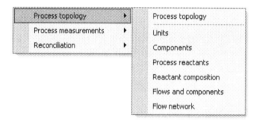

Figure 12.5: Example of a navigation menu

Adding navigation menus   You can add a navigation menu to any menu in the **Menu Builder** tool (see Section 12.3). For each navigation menu you must specify a reference page and the scope of the subtree to be displayed in a similar fashion to that illustrated for the navigation object in Figure 12.4.

Hiding pages   Pages can be statically or dynamically hidden using the page **Properties** dialog box (see also Section 11.2), as illustrated in Figure 12.6. In the **Hidden** field, you must enter a scalar value, identifier or identifier slice. Whenever the property assumes a nonzero value the page is hidden, and automatically removed from any navigational interface component in which it would otherwise be included.

Authorizing access   For larger applications, end-users can usually be divided into groups of users with different levels of authorization within the application. Disabling pages based on the level of authorization of the user (explained in Chapter 20) then provides a perfect means of preventing users from accessing those data to which they should not have access. You can still open a hidden page via a hard-coded page link.

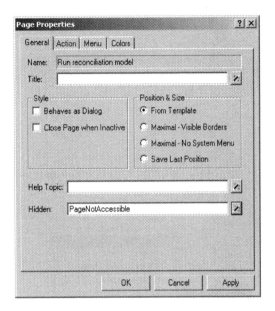

Figure 12.6: Hiding a page

12.2 The Template Manager

Complementary to the **Page Manager** is the AIMMS **Template Manager**. Using the **Template Manager**, you can ensure that all pages are the same size and possess the same look-and-feel, simply by positioning all end-user pages in the template tree associated with a project. An example of a template tree containing both templates and end-user pages is displayed in Figure 12.7.

Consistent look-and-feel

In addition to all the end-user pages, the template tree can contain a hierarchical structure of template pages. Within the template tree, template pages behave as ordinary pages, but they are not available to end-users. Through templates you can define common page objects that are shared by all the template *and* end-user pages positioned below a particular template in the template tree.

Hierarchical template structure

When you want to use the same template page at two or more distinct positions in the template tree, AIMMS lets you duplicate, rather than copy, the template node containing that component. Changes made to the duplicated page template at any position in the template tree, are automatically propagated to all other occurrences. Duplicated templates can be recognized by the duplication symbol ⌸ which is added to the icon of every duplicate template in the template tree.

Duplicating page templates

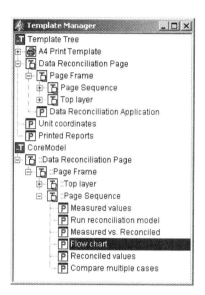

Figure 12.7: The **Template Manager**

End-user pages automatically added

Every new end-user page created in the **Page Manager**, is automatically added to the root node in the template tree. By moving the page around in the template tree, it will inherit the combined look-and-feel of all templates above it.

Common page components

The hierarchical structure of the template tree lets you define layers of common objects on top of each other. Thus, a first template might globally define the page size and background color of all underlying pages, while a second template could define common components such as a uniformly shaped header and footer areas. As an example, Figure 12.8 illustrates a template for an end-user page from the template tree of Figure 12.7, in which the components defined in various templates are identified.

Modify look-and-feel

You can quickly modify the entire look-and-feel of your application, by moving a subtree of templates and end-user pages from one node in the template tree to another. Thus, the entire look-and-feel of page size, header and footer areas, background color and navigational area(s) of all pages in an AIMMS application can be changed by a single action.

Template objects not editable

When you open a template or end-user page in the template manager, it will be opened in edit mode by default, and inherit all the properties of, and all objects contained in, the templates above. On any template or end-user page you can only modify those objects or properties that are defined on the page itself. To modify objects defined on a template, you must go to that template and modify the objects there.

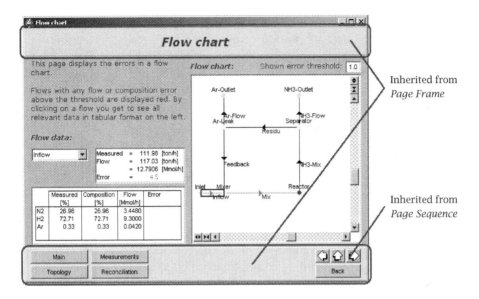

Figure 12.8: Example of an end-user page using templates

You can achieve an exceptionally powerful combination by adding navigational components to a template page. If the reference page property of such a navigational component is expressed in terms of the current page, or one of its ancestor pages, then, in end-user mode, the current page will always refer to the particular end-user page which uses that template. Thus, given a well-structured page tree, you potentially only need a *single* template to add navigational control components to *all* end-user pages. This is particularly true for such common controls as **Previous** and **Next** buttons.

Combine with navigational components

12.2.1 Templates in library projects

Each library project in AIMMS has a separate template tree, which is available as a separate root node in the **Template Manager**, as illustrated in Figure 12.7. Pages in a library must be positioned in the template tree of that library to obtain their look-and-feel. This allows the developer of a library project to define the look-and-feel of the pages in the library completely independent of the main project and other library projects.

Templates in libraries

If you want the pages of an entire application to share a common look-and-feel across all library projects included in the application, AIMMS also allows you to duplicate template pages from the main project into a library project. Thus, any changes made to the templates in the main project are automatically inherited by the pages in the library that depend on the duplicates of these templates.

Sharing templates with the main project

Sharing templates across multiple projects

Conversely, you can also use library projects to enable the end-user GUIs of multiple AIMMS project to share a common look-and-feel. By defining templates, fonts, and colors in a single library project, and including this library project into multiple AIMMS projects, the pages in these projects can depend on a single, shared, collection of page templates. Thus, changes in a single template library will propagate to all AIMMS projects that depend on it.

Moving pages to a library

If you move or copy pages from the main project to a library project (or between library projects), AIMMS will automatically duplicate the relevant template structure from the source project to the destination project. This ensures that the pages have the exact same look-and-feel at the destination location as they had at their source location.

Example

The automatic duplication behavior of AIMMS is illustrated by the page tree in Figure 12.1 and the template tree in Figure 12.7. These trees were created by moving the *Reconciliation* page and its child pages from the main project to the *CoreModel* library. Subsequently, AIMMS automatically duplicated the template structure in the *CoreModel* library to ensure the identical look-and-feel for the moved pages.

12.3 The Menu Builder

The Menu Builder

The last page-related design tool available in AIMMS is the **Menu Builder**. With the **Menu Builder** you can create customized menu bars, pop-up menus and toolbars that can be linked to either template pages or end-user pages in your application. The **Menu Builder** window is illustrated in Figure 12.9. In the **Menu Builder** window you can define menus and toolbars in a tree-like structure in a similar fashion to the other page-related tools. The menu tree closely resembles the natural hierarchical structure of menus, submenus and menu items.

Default menu bar and toolbar

As illustrated in Figure 12.9, the **Menu Builder** will always display two nodes representing the standard end-user menu bar and toolbar. These bars are linked to all end-user pages by default. Although non-editable, you can use these nodes to copy (or duplicate) standard end-user menus or submenus into your own customized menu bars and toolbars.

Inserting new nodes

In the *User Menu Tree*, you can add nodes to represent menu bars, (sub)menus, menu items or toolbars in exactly the same manner as in other trees such as the model and page trees. Also, you can copy, duplicate or move existing nodes within the tree in the usual manner (see Section 4.3). The names given to menu and menu item nodes are the names that will be displayed in the end-user menus, unless you have provided a model-specific menu description in the menu **Properties** dialog box (e.g. to support multiple languages).

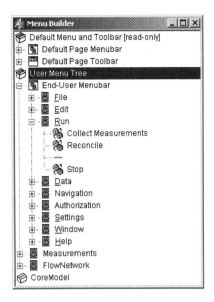

Figure 12.9: The **Menu Builder** window

For every node in the menu tree you can modify its properties through the **Properties** dialog box. In the **Properties** dialog box you can perform tasks such as linking end-user actions or model procedures to a menu item, provide shortcut keys, tooltips and help, or link a menu item to model identifiers that specify whether the item should be disabled within an end-user menu, or even be completely hidden from it. The **Properties** dialog box for a menu item is shown in Figure 12.10.

Menu item properties

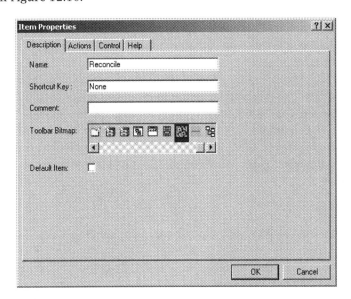

Figure 12.10: The menu item **Properties** dialog box

Adding menu actions

Through the **Actions** tab of the **Properties** dialog box, you can associate a list of actions with a menu item. Such actions can consist of executing menu items from system menus, navigational commands such as opening or closing pages, and also running procedures from your model, verifying assertions or updating identifiers.

Hiding and disabling items

With the **Control** tab it is possible to provide control over a menu item from within your model. You can specify scalar 0-1 identifiers from within your model to determine whether a menu item or submenu should be disabled (grayed out) or completely hidden from the menu. Thus, you can prevent an end-user from performing tasks for which he is not authorized. In addition, you can couple a 0-1 identifier to a menu item in order to determine whether a menu item is checked, and which conversely toggles its value when an end-user checks or unchecks the item.

Tooltips and help

In the **Help** tab of the **Properties** dialog box, you can provide a description and help describing the functionality of a menu command. It lets you specify such things as the tooltips to be displayed for buttons on the button bar, a descriptive text for to be shown in the status bar, and a link to a help item in the project related help file.

Navigation menus

Navigation menus are a special type of menu that can be added to the menu tree. Navigation menus expand to a number of items in the current menu, or to one or more submenus, according to the structure of a particular subtree of the page tree as specified by you. Through navigation menus you can quickly and easily create menus that help an end-user navigate through your application. For example, you could create a menu item which links to the first child page, or to the parent page, of any page to which the menu is linked. The details of how to specify which pages are displayed in a navigation menu can be found in Section 12.1.2.

Linking to pages and objects

You can link a single menu bar, toolbar and pop-up menu to any end-user or template page in your project through the **Menu** tab of the page **Properties** dialog box, as illustrated in Figure 12.11 For every field in the dialog box, AIMMS lets you select an existing node in the menu tree. If you do not specify a menu bar or toolbar, AIMMS will automatically open the default end-user menu bar and toolbar.

Inherited menus

When you add a menu bar or toolbar to a page template, these bars are automatically inherited by all pages that use that template. In this manner, you can quickly add your own customized end-user menu to all, or groups of, pages in your application. All new end-user pages will, by default, inherit their menu bar and toolbar from their templates.

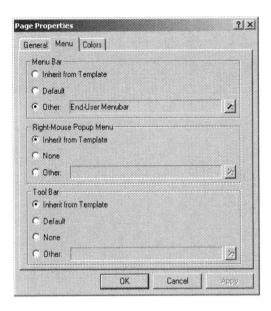

Figure 12.11: Linking menus to pages

12.3.1   Menus in library projects

In addition to the main *User Menu Tree* in the **Menu Builder**, each library project in AIMMS has a separate menu tree, as illustrated in Figure 12.9. In this menu tree, you can create the menus and toolbars that are specific for the pages defined in the library at hand.

Menus in library projects

When you are specifying the actions associated with a menu item in the menu tree of a library, you have access to all the identifiers and procedures declared in the library module of that library. For menu items in all other menu trees, you can only access the identifiers in the interface of the library.

Accessing private identifiers

When creating menus and toolbars in a library, you can duplicate menus and menu items from any other menu tree in the **Menu Builder**. Likewise, you can duplicate menus and menu items from a library project into the menu tree of the main project. This enables you to compose global menu- and toolbars that can be used in the overall application, yet containing library-specific submenus and menu items to dispatch specific tasks to the appropriate libraries.

Creating menus

When you want to assign a menu to a page or template, AIMMS allows you to choose a user menu of either the main project or of any of its library projects. You should note, however, that choosing a menu from a different library project creates an implicit dependency on that project which is not immediately apparent in the page or template tree. If you copy or move pages

Assigning menus to pages

with a user menu or toolbar from one project to another, AIMMS will not duplicate that menu or toolbar, but still refer to their original locations as expected.

Chapter 13

Page Resizability

Due to the diversity of objects and their position on a page, it is not immediately clear how objects should adjust when the size of a page is changed. Should buttons remain the same, when the size of particular data objects are changed? Such decisions are up to you, the developer of the application.

Resizability

In this chapter, you will learn about the facilities in AIMMS which you can use to specify how page components should scale when a page size changes. Such facilities allow you to create *resizable pages* which are ready for use with different screen resolutions. In addition, resizable pages let an end-user temporarily enlarge or reduce the size of a particular page to view more data on the same page, or to simultaneously look at data on another end-user page.

This chapter

13.1 Page resizability

When you are developing an end-user interface around an AIMMS-based application for a large group of end-users, you must decide about the base screen resolution on which the end-user interface is intended to be run primarily. Such a decision is based on your expectations about the screen resolution that most of your end-user will be using. Fortunately, there is a tendency towards high-resolution screens amongst users.

Choosing a base resolution

Nevertheless, it is likely that one or more of your end-users will request to run the application at a different resolution. One reason could be that they use a notebook which does not support the base resolution you selected. Another reason could be that some of your end-users are working with such large data sets that a higher resolution would help them to have a better overview of their data.

Supporting different resolutions

To help you support the demands of your end-users, AIMMS provides a fairly straightforward facility to create resizable pages and page templates. As you will see later on, the position and type of so-called *split lines*, placed on a resizable page, determines the manner in which objects on the page will scale upon resizing the page.

Resizable pages

Non-resizable behavior

When a page has not been made resizable, all objects on that page will remain in their original position. Whenever such a page is reduced, and a data object included on the page falls outside the visible page area, AIMMS will automatically add horizontal or vertical scroll bars. If the page is increased in size, the bottom and right parts of the page will remain empty.

Adding resizability

To make a page resizable, the page should be put into **Resize Edit** mode, which is available in the **View-Resize Edit** menu of any page that is already in **Edit** mode. **Resize Edit** mode will replace all objects on the page by shaded rectangles. Figure 13.1 illustrates the **Resize Edit** view of the end-user page shown in Figure 12.8.

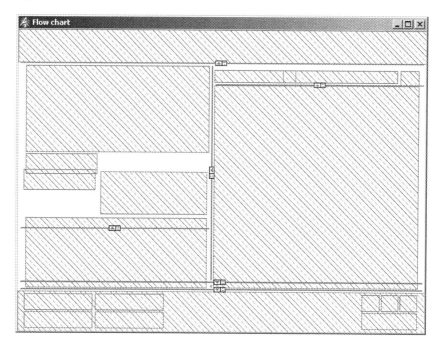

Figure 13.1: Page 12.8 in **Resize Edit** Mode

Split lines

A page is resizable as soon as it contains one or more horizontal or vertical *split lines*. The page in Figure 13.1 illustrates a number of such split lines. Each split line divides the rectangle in which it has been placed into two subrectangles. For each vertical split line you can decide either

- to keep the width of the left or right subrectangle constant (indicated by ▣ and ▣ markers),
- to ensure that the widths of the left and right subrectangles have the same ratio (indicated by ▮ marker), or
- to make the ratio between these two widths user-adjustable (indicated by ▣ or ▣ markers).

Similarly, horizontal split lines are used to indicate the relative height of the lower and upper subrectangles. On an end-user page, a user-adjustable split line will be visible as a split bar, which can be dragged to (simultaneously) resize the areas on both sides of the bar.

By selecting a subrectangle created by a split line, you can recursively subdivide that rectangle into further subrectangles using either horizontal or vertical split lines. What results is a specification of how every part of the page will behave relative to its surrounding rectangles if the size of the entire page is changed.

Stacking split lines

One way of adding split lines to any subrectangle on a page in **Page Resize** mode is to select that subrectangle on the page (by clicking on it), and add a horizontal or vertical split line to it using one of the buttons 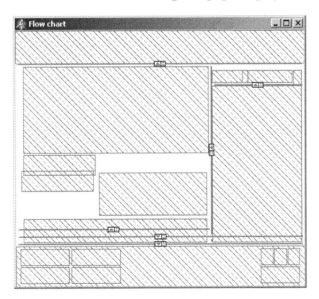 from the **Page Resize** toolbar. Alternatively, if you want to insert a split line within an existing hierarchy of split lines, select the line just above where you want to insert a split line, and use one of the buttons to insert a new split line of the desired type.

Adding split lines

By putting a page that is already in **Resize Edit** mode into **Resize Try** mode (via the **View-Resize Try** menu) and resizing the page, AIMMS will display the shapes of all page objects according to the specified resize behavior. Figure 13.2 illustrates the effect of resizing the page displayed in Figure 13.1

Resize Try mode

Figure 13.2: Resizing a resizable page

to a smaller size. These resized shapes are determined by calculating the new relative positions of all four corner points of an object within their respective surrounding rectangles. This may result in nonrectangular shapes for some page objects, which are marked red. In such a situation, you should reconsider the placement of objects and split lines. Non- rectangularly shaped objects may distort the spacing between objects in end-user mode, because AIMMS will enforce rectangular shapes in end-user mode by only considering the top-left and bottom-right corners of every object.

Example   Consider the configuration of split lines illustrated in Figure 13.1, and its associated end-user page displayed in Figure 13.3. As already indicated in Fig-

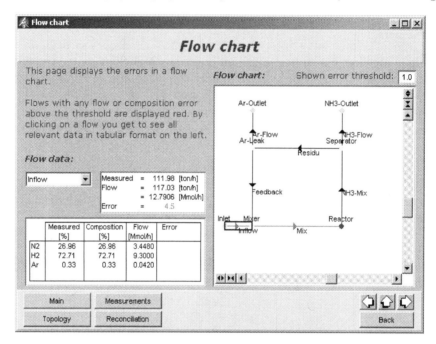

Figure 13.3: End-user page associated with Figure 13.1

ure 13.2, the particular combination of split lines results in the following behavior.

- The header area will have a fixed height at the top of the page whatever the page height, but will grow (or shrink) along with the page width.
- Similarly, the entire footer area will remain a fixed distance from the bottom of the page, and grow along with the page width.
- The information on the left-hand side of the data area has a fixed width, and the table will only grow/shrink vertically along with the page height.
- The flow chart header on the right-hand side of the data area has a fixed height, while the flow chart itself will grow/shrink along with both the page height and width.

When entering **Edit** mode, AIMMS will always restore the editable page area to its original size (as saved at page creation time). This ensures that objects placed on the page always use the same coordinate system, preventing pixel rounding problems during a page resize. If the page has been saved at a different end-user size, AIMMS will open the page frame at the latest end- user size, and make the parts outside the original (editable) page size unavailable for editing, as illustrated in Figure 13.4. Any split line added to a page (or to its templates), will be visible in a page in **Edit** mode as a thin line.

Original size only

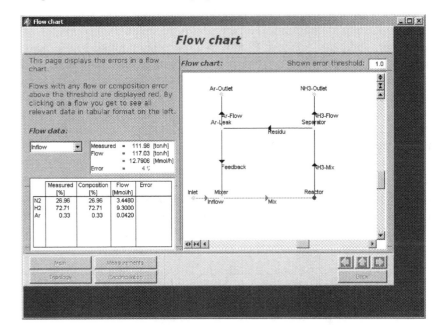

Figure 13.4: Editable area and split lines of a resizable page in **Edit** mode

13.2 Resizable templates

When you are creating an AIMMS-based application with many resizable pages, all based on a number of page templates, you should also consider defining the basic resize properties of these templates. As templates behave as ordinary pages in the template tree, you can add split lines to templates as described in the previous section.

Creating resizable templates

All templates and end-user pages based upon a resizable template inherit the resize properties of that template, i.e. all split lines in the template are also applicable to its child templates and pages. Generally, such inherited split lines should take care of the resize properties of those objects that are contained in the template itself.

Inherited resizability

Adding split lines   On any page (either template or end-user page) you can always add additional split lines to those inherited from its ancestor template(s). The added split lines are used to specify the resize properties of the additional objects that have been placed on the page. In this manner, the template tree can be used to define the entire look-and-feel of your pages in a hierarchical manner, and their resize properties.

Example revisited   The example page in Figures 13.1 and 13.3 already illustrates the inherited resizability from templates. In fact, Figure 13.1 displays the split line configuration of a template defining the common header and footer area of all its child pages. The page in Figure 13.3, which uses this template, automatically inherits its resize properties. Therefore, the table in the "data area" of this page automatically grows or shrinks in relation to the page size as dictated by the template.

13.3   Adapting to changing screen resolutions

Coping with different resolutions   AIMMS allows you to create pages in such a manner that they will automatically adapt to changing screen resolutions. Thus, given a sensible configuration of split lines, you can create an application than can be run in resolutions other than the base resolution for which you developed the pages.

Page properties   To specify the behavior of pages and templates, open the **Properties** dialog box for the page (template), as illustrated in Figure 13.5. In the **Position & Size**

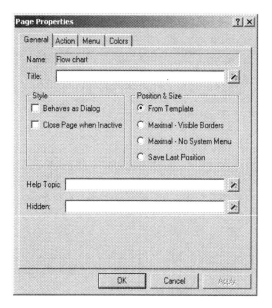

Figure 13.5: The page **Properties** dialog box

area of this dialog box, you can select the default position and size of the page, which AIMMS will use when opening the page.

For every page in your application, you can select one of the four following standard page opening modes:

Opening modes

- get the position and size from the template used by the page,
- open the page at maximum size, but with visible page borders,
- open the page at maximum size, but without visible page borders, and
- open the page using the last saved position and size.

If you specify that a page should obtain its position and page size from its template, the page will use the page open mode as specified for that template. When, in turn, this template has been specified to open according to its last saved position and size, an interesting interaction between the template and all its dependent pages will take place. Changing the position and size of *any* page using such a template will cause all the other pages using that template to be opened using the new position and size.

Inherited modes

As an application for the above, you could decide to make every page and page template dependent on the position and size of the root template. In this manner, changing the size of any page, will automatically result in the adjustment of every other page.

Resizable root template

When you have specified that a page or page template should save its last position, this position is stored between sessions. That is, the next time you open the same project, AIMMS will open such pages in the same positions as used in the previous sessions on the same computer.

Save over sessions

Chapter 14

Creating Printed Reports

This chapter

Besides an attractive graphical end-user interface, paper reports containing the main model results are also an indispensable part of any successful modeling application. This chapter details printed reports. Printed reports are created and designed in a similar fashion to ordinary end-user pages, and can contain the same graphical objects for displaying data. There is, however, additional support for dividing large objects over multiple printed pages.

14.1 Print templates and pages

Printing versus GUI

AIMMS makes a distinction between *end-user* pages that are designed for interactive use by the end-user of your application and *print* pages that are specifically designed for printing on paper. While this may seem puzzling at first, a closer inspection reveals a number of serious drawbacks associated with printing ordinary end-user pages. The most important are:

- usually the screen resolution does not match the size of a sheet of paper,
- in a printed report, you cannot rely on the use of scroll bars on either the page itself or within objects if all the available information does not fit,
- the use of background colors may look nice on the screen, but often severely hinders the readability of printed reports, and
- you may want to add header and footer information or page numbers to printed pages, which are not part of an end-user page.

Printing ordinary end-user pages

Through the **File-Print** menu, AIMMS allows you to print a simple screen dump of the contents of any end-user page that currently is on the screen in your application. The **File-Print** menu will open the **Print Page** dialog box illustrated in Figure 14.1. Using this dialog box you can choose the size, border width and orientation of the screen dump to be produced. Any data that is not visible on the end-user page will also not appear in the screen dump.

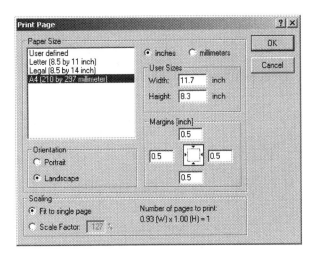

Figure 14.1: **Print Page** dialog box for end-user pages

An AIMMS print page, on the other hand, allows you to compose a customized report with data objects that can automatically be resized to print all available object data. Print pages are characterized by the fact that they depend on a special *print template* in the template tree. You can add a print template via the **New-Print Template** item in the **Edit** menu of the **Template Manager**. Print templates can only be placed at the top level of the template tree, i.e. directly below the root, as illustrated in Figure 14.2. All pages below the print template behave as print pages.

Print templates and pages

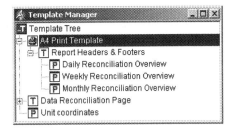

Figure 14.2: Example of a print template

Every print template has an associated paper type. The paper type lets you define properties such as paper size, paper orientation, and the width of the surrounding margins. By default, AIMMS will create new print templates with the predefined A4 paper type. You can modify the paper type by opening the print template and selecting **Paper Type** in the **View** menu, which will then open the dialog box displayed in Figure 14.3. With it, you can either select one of the predefined paper types, or define a custom paper type by specifying the paper size, orientation and margins yourself.

Specifying paper type

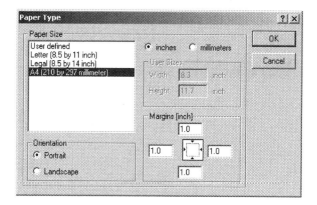

Figure 14.3: **Paper Type** dialog box

Page border

When you open a print page (or print template) in edit mode, AIMMS displays a rectangular box representing the margins corresponding to the current paper type. An example of an empty print page in landscape format containing a margin box is illustrated in Figure 14.4. The margin lines are not displayed

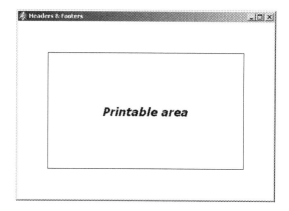

Figure 14.4: An empty print page in landscape format

when the page is previewed or printed. In edit mode, however, the margin lines may help you to position data objects within the printable area.

Printing pages
with margins

In general, AIMMS will print all objects on a print page, regardless of their placement with respect to the page margins. However, when you have indicated that a data object should be printed over multiple pages (as discussed in the next section), AIMMS will always restrict itself to printing within the indicated page margins.

You can add data objects and graphical objects to a print page in exactly the same way as you can add such objects to an ordinary end-user page. In fact, objects contained on your end-user pages which you want to be part of a printed report as well, can be copied directly to a print page by means of a simple copy and paste action.

Adding objects

You should note, however, that not all objects that are placed on a print page (and are visible on the screen) will be printed on paper. Specifically, AIMMS will omit all controls such as buttons and drop-down lists which are intended for interactive use only. Through such controls you can provide special facilities for your end-users such as allowing them to make last-minute choices prior to printing, activate the actual print job, or navigate to a previous or next print page through navigational controls linked to the page tree (see Section 12.1.2). To prevent interference with printable objects, non-printable controls are best placed in the page margins. Naturally, you can also place such controls on a separate dialog page.

Non-printable objects

You can add one or more normal templates below any print template in the template tree, in exactly the same way as for ordinary end-user pages (see also Section 12.2). In this way, you can specify common components such as headers and footers, that are automatically inherited by all dependent print pages.

Using additional templates

Page numbers can be added to a print page by displaying the predefined AIMMS identifier CurrentPageNumber either on the page itself or on any of its page templates. When printing a single page, AIMMS resets CurrentPageNumber to 1, and will number consecutively for any additional pages that are created because of a large data object. When printing a report that consists of multiple print pages (see below), AIMMS resets CurrentPageNumber to 1 prior to printing the report, and increments it for every printed page.

Displaying the page number

AIMMS allows you to print a print page in several manners:

Printing print pages

- when the page is opened on the screen, you can print it using the **File-Print** menu,
- you can attach the above action to a page or toolbar button, by adding the **File-Print** menu action to the button, or
- you can print the page from within the model using the PrintPage function.

In addition to printing single print pages, AIMMS also allows you to print entire reports consisting of multiple print pages. Printing such reports can be initiated only from within your model, through calls to the predefined functions PrintStartReport, PrintPage and PrintEndReport. A multipage report is started by a call to PrintStartReport, and finished by a call to PrintEndReport. All the

Printing complete reports

single print pages constituting the report must be printed through consecutive calls to the PrintPage function in between. Such a composite report will be sent to the printer as a single print job, and by default all pages within the report will be numbered consecutively starting from 1. However, if you so desire, AIMMS allows you to modify the value of CurrentPageNumber between two consecutive calls to the PrintPage function. The print functions in AIMMS are discussed in more detail in Section 18.4.2.

14.2   Printing large objects over multiple pages

Printing large data objects

Print pages are explicitly designed to allow the printing of data objects that hold large amounts of data and, therefore, do not fit onto a single page. On a print page you can specify that such large objects should be split over as many pages as are needed to display all the underlying data, respecting the specified page margins. In addition, AIMMS allows you, in a flexible manner, to further restrict printing to those parts of the print page that are not already occupied by fixed page components such as headers and footers.

Required steps

In order to enable multipage printing, only two simple steps are required. More specifically, you should

- modify the print properties of both the fixed page components and the large data objects contained on the page to specify their desired printing behavior, and
- create a simple subdivision of the print page by means of the resize split lines (see also Chapter 13) to specify how objects should be fixed to particular page positions, or resized as necessary.

The remainder of this section discusses both steps in more detail, and illustrates them on the basis of a realistic example.

Specify printing occurrence

For every object on a print page or template you can define when and how the object should be printed. Through the **Misc** tab of the object **Properties** dialog box (as displayed in Figure 14.5) you can specify that an object must be

- printed on every printed page (such as headers or footers),
- printed on only the first page or the last page,
- printed on all pages except for the first or the last page (ideal for indicating whether the report is continued or not),
- spread over as many pages as required to display all its associated data, or
- omitted from the printed output.

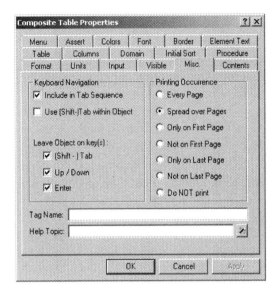

Figure 14.5: The **Misc** properties tab

Using these choices, you have the capability of having a single print page printed over multiple pages where each page component behaves as desired. For instance, headers and footers can be printed on every page or, perhaps, on all pages but the first. A report title needs only be displayed on the first page. Data objects which are expected to become large can be split over multiple pages.

Usage

By default, all objects will be printed the same size as they have been placed onto the print page during its creation. Thus, without further action, a large table is split over multiple pages based on the original table size. As you will see below, objects can be stretched to fill the entire print page by using AIMMS' resizability features.

Multiple page printing

Two types of split lines are useful when creating a resizable template for a printed report. *Fixed distance* split lines can be used to specify those areas of a page that contain components such as headers and footers which should keep their original shape and position. *Adjustable distance* split lines can be used to indicate that the objects contained in the relevant page area must be adapted to fill the maximum space within that area.

Resizing page objects

Whenever a data object does not fit in an adjustable area, AIMMS will first extend the data object to the border of the adjustable area. This border may be either the page margin, or a fixed distance split line that has been placed on the page. When AIMMS runs into the border of an adjustable area, further printing of the data will continue on a new page. On the final page, AIMMS will

Spreading over multiple pages

reserve just enough space to contain the remaining data.

Multiple splits   By creating multiple adjustable areas just below or alongside each other, you have the opportunity to place multiple data objects of varying size within a single report, with each object in its own resizable area. Once AIMMS has finished printing the object contained in the first resizable area, it will start printing the next object directly adjacent to the first, in either a left- to-right or top-to-bottom fashion, depending on your selected layout.

Use of templates   If you are creating multiple reports with more or less the same layout of headers and footers, you should preferably use template pages to define such fixed page components, together with their appropriate printing occurrence (e.g. first page only) and resizability properties for multiple page printing. If you use such templates wisely, creating a specific print page boils down to nothing more than adding one or more data objects to the data area of the page (i.e. the page area not occupied by a header and/or footer), and defining the appropriate print and resizability properties.

Example   The print page, and its corresponding configuration of split lines contained in Figure 14.6 illustrate AIMMS' capabilities of printing large data objects over multiple pages. In this page, the header and footer components are enclosed

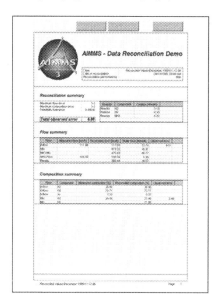
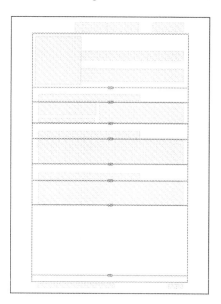

Figure 14.6: Print page with in edit and resize edit mode

in areas which have a fixed size from the top and bottom margin, respectively, and are printed on every page of the report. The middle part of the page contains a number of data objects, each enclosed in an adjustable area from the top down. As a result, AIMMS will split each object subsequently over as many pages as are necessary. Figure 14.7 illustrates a multipage report generated from this print page for a particular case. Note that the navigation and print buttons that are placed outside the page margins of the print page for use during page preview (see also Section 14.1), are excluded from the actual report that is printed.

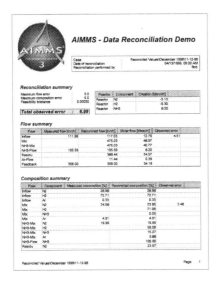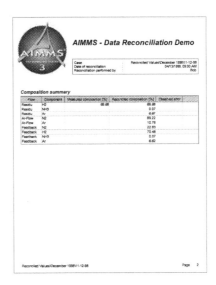

Figure 14.7: Example of a multipage report

Chapter 15

Designing End-User Interfaces

This chapter

The goal of this chapter is to give some directions that will help you design an end-user interface that is both easy to use and straightforward to maintain. Even though design is an intuitive subject, you may find that developing a professional interface is no trivial matter. The general design principles outlined in this chapter, as well as the tools provided by AIMMS (e.g. the **Template Manager**, the **Page Manager** and the page design tools), will help you to specify and maintain a high-quality model-based interactive report.

Linear design is not recommended

A linear design process consists of the following steps.

- Develop an extensive technical specification of the interface following consultation with prospective users.
- Let the prospective users read and evaluate the specification.
- Adjust the specification and implement the interface accordingly.

Even though this approach may seem natural, in practice it does not work well. Users are usually unable to specify precisely what they require, and they find it difficult to imagine the look and feel of a graphical user interface from a written document. That is why we recommend an iterative design process based on prototyping.

Design through prototyping recommended

Rather than writing a detailed technical specification of a graphical user interface you can construct an initial design immediately following a consultation with prospective users. You will find that the evaluation of a well-designed prototype is a much better way of helping users structure their wishes, and it provides them with an impulse for new ideas. With AIMMS' point- and-click tools the actual construction of, and subsequent adjustments to, your interface are not a major task. As a result, you will be able to complete an interface in a limited number of iterations with the assurance that it will be accepted by your final users.

Main subjects

This chapter focuses on three major topics related to interface design, each of which is covered in a separate section. These are

- page design,
- navigation between pages, and

■ quality of interaction with your end-user.

In the first section of this chapter these three topics are briefly introduced with a quick orientation and summary. In the final section you will find some pages taken from real-life applications in order to illustrate the principles discussed in this chapter.

15.1 Quick orientation

This section gives you a quick and pragmatic overview of some of the considerations you should take into account when designing an end-user interface.

This section

If you design an interface for yourself, your only concern is to improve your own efficiency while developing and debugging your AIMMS model. In this case the following guidelines are relevant.

Page design for yourself

■ Design one or more single-page overviews combining related input and output data on the same page.
■ Use a template page to setup and link the different data pages.
■ Link procedures that perform the various data manipulations required by your model using buttons on the relevant pages.

Should you be designing the interface for someone else, then your basic concern is to make sure that your design will be used effectively. The key factor here is the quality of communication. The interface should possess the following characteristics:

Page design for someone else

■ a pleasant look and feel (all pages use the same layout and all objects have clear descriptions in their title and status line),
■ a consistent navigational structure (making the behavior of the system predictable), and
■ robustness (through clear error messages and extensive error checking).

These characteristics should lead to user-friendliness and thereby to acceptance of the interface by your end-users.

The number of pages in an application will depend upon the size and complexity of the model, as well as on the number of options you want to provide to your end-users. In an application with few pages the navigation structure can be quite simple. For example, a wheel-like structure linking all pages will suffice. In such a structure each page is linked to the previous, the next, and the main page, with the main page providing direct access to all other pages.

Navigation with few pages

Navigation with many pages

If the application consists of a large number of pages, then a wheel-like structure with a single main page is not practical. In this case, a tree structure with cross links between the pages is a good option to facilitate ease of navigation. The following characteristics contribute directly to the success of such a structure:

- main sections that are easily accessible and subsections that contain their own outlines or menus,
- section headings in a fixed position on every page,
- a customized menu bar that can be used to give quick access to important pages as well as to provide menu items for general actions that can be executed from any page (for instance a print command), and
- extra orientation clues by associating colors (for instance in the title bar) with the different sections.

Interaction with occasional users

Occasional users will not require much control over the behavior of the model. They view the interface as an easy way of browsing through information, and occasionally carrying out some experiments. An appropriate interface should encompass the following characteristics:

- easy to read with clear explanations attached to symbols and icons,
- summaries of important model results,
- graphs for quick trend perception,
- no advanced control options requiring explanation, and
- a tree structure of pages enabling occasional users to quickly find information.

Interaction with frequent users

Frequent users usually know a lot about the underlying application. They tend to use the application as an operational tool, and are prepared to spend some time learning how to use it. This will allow you to build more advanced functionality into the interface. In addition to those characteristics mentioned above, we suggest that the following characteristics should be included:

- advanced control options for quick access throughout the interface,
- page design adjusted to familiar report formats,
- use of existing color conventions,
- a wide tree structure with relatively few levels,
- a setup to simplify the import of new data, and
- a help system, or a number of separate pages displaying help text.

What's next

In the next three sections the subjects of *page design, navigation,* and *interaction* will be discussed in more detail. These sections, together with the summary provided in this section, form a basis on which you should be able to design a high-quality interface for your end-users.

15.2   Page design

First, we will provide a few general guidelines for page design. Some of them will be elaborated in the subsequent paragraphs.

General page design principles

- All pages should use the same layout. Important buttons appearing on every page should always appear in the same place. The easiest way of achieving this is to use page templates.
- Give pages a clear title. The title is often the user's first clue as to the contents of the page.
- In order to clarify the meaning of each page place comments in the status line, title or descriptive text of each page object. Additionally, do not put too much text on a page.
- Limit the use of colors for titles and areas to a balanced combination of two to four colors.
- Use only those fonts that are installed on all computers, and pay attention to their readability.

The layout of pages throughout an interactive report will become more consistent if you divide each page into different areas each containing a group of related objects. There are several areas that you could consider:

Divide page into areas

- the page title and section indicator,
- the data object(s),
- navigational buttons (previous, next, main, go back),
- buttons with actions such as checks and calculations,
- a logo,
- one or more floating indices, and
- a reference to a certain model state, currently loaded data set, etc.

Visually, you can use borders and/or colors to highlight these areas. Two typical examples of how to divide a page are given in Figure 15.1.

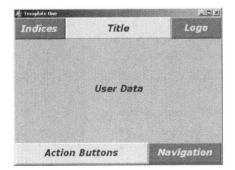 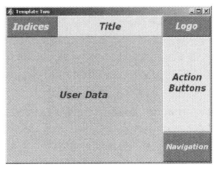

Figure 15.1: Example page layouts

Bitmap buttons   By using bitmaps on buttons users can quickly recognize their meaning. For example, most people will interpret an arrow on a button faster than the word 'Next'. Sometimes the combination of an icon and a text is preferred. Initially, users will tend to read the text to identify the button's action, while later on just the icon will suffice. You should ensure that both the style and the size of the bitmap used in your interface are consistent, and that you do not clutter the interface with too many different bitmaps.

Color is   Experience has shown that the use of color is crucial in the acceptance of your
important   interface. The golden rule is to be sensitive to the wishes of the end-users, and to use combinations of colors that most people will appreciate (not an easy task). Suggestions on your part could reflect such items as house style, logo, or colors that everyone automatically associates with the underlying application.

Some guidelines   Some guidelines concerning the use of color on a single page are as follows.

- Colors can be used as a way to visually segment the page into separate regions, or to draw the user's attention to a particular point.
- Even though you may be tempted to use lots of different colors, it is wise not to do so. Too many colors will clutter the screen and tire the eyes.
- The contrast between foreground and background colors must be sufficient to facilitate reading.
- Selected colors should not conflict with familiar interpretations.
- Color choice should always be such that color-blind users are able to use the interface without problem.

Decide on a   A predetermined coloring palette will give each page a consistent look. You
coloring palette will need colors for *page format*, *page text*, and *page highlights*.

- Format colors make up the page backgrounds. They are colored rectangles behind data objects, text and logos. Light gray is frequently used as a background color, because buttons look good and shadow effects come up nicely.
- Text colors are used for titles, foreground color in data objects, etc. There should be a clear contrast with the format colors.
- Highlight colors are used for the remaining objects. In AIMMS you can specify your own color schemes and link these to particular data objects.

User-adjustable   If your application is used by many users, it may be impossible to satisfy the
colors   color preferences of all users. In that case, you can define all colors in the interface through color parameters, and create a color setup page in which the user can select his preferred color scheme.

The size and style of the fonts will directly affect the look and readability of each page. Just as with colors you should avoid using too many different fonts, as they will give the interface a disorganized look.

Fonts are also important

Here are some extra points you may wish to consider when selecting fonts for objects on a page.

Some guidelines

- Generally, sans serif fonts (such as Arial) are more readable on a computer screen than fonts with serifs (i.e. fonts with small wedge-shaped extensions to each character such as Times). This is particularly true for small font sizes.
- Similarly, regular fonts are more readable on a computer screen than italic fonts.
- Words in (the familiar) lower case are easier to read than words in capitals. Words in upper case can be used to attract attention as long as they are used sparingly.
- Vary the size of characters for emphasis, but try to limit the number of sizes. In this way a user will recognize the implied hierarchical structure behind the text.

AIMMS allows you to put together your own font list, and give names to the fonts. You can take advantage of this facility by naming fonts in terms of their functionality. Typical examples are fonts named "page title", "button" and "table". This will help you to make consistent font choices for each object during page construction. Should you subsequently decide to change the fonts during the maintenance phase of your interface, then all you need to do is to edit the font in the font list, and all objects with that font name will be automatically updated.

Give fonts functional names

The number of colors supported by the video display adaptor determines the possibilities in using colors. The screen resolution of the monitor determines the size and contents of pages. Not all available fonts are installed on every computer. It is wise to take these technical limitations into account by checking the hardware limitations of your end-users. You could always use fewer shades of color, design for lower resolution, and limit your choice to standard fonts.

Be aware of technical limitations

15.3 Navigation

An AIMMS application basically consists of pages that are linked by buttons. These pages should be presented in an order that is both meaningful and logical to the end-user. This is where navigation becomes important. The AIMMS **Page Manager** helps you with navigation. The following general guidelines may be helpful.

General navigation principles

■ The performance of the system should be predictable. A user will create a mental picture based on his experience with the current, and similar, systems. Try to adhere to standards set by other systems.

■ A user should always be able to return to the page he just left. AIMMS offers a specific button action for this purpose.

■ Give the user easy access to page overviews from buttons placed on every page, or via submenus (accessible from every page).

■ When the number of pages is small, use a wheel structure to navigate. All pages are then linked through buttons to the previous and the next page, as well as to a single main page from which all other pages are accessible.

■ When the number of pages is large, use a tree structure to navigate. Then the number of steps needed to arrive at any particular page is at most the number of levels in the tree. The wheel structure can still be used for small self-contained subsets of pages.

Navigate between sections

When linking pages to improve navigation throughout the interface, it helps to distinguish sections of pages that belong together. Typical sections are:

■ *input sections* enabling an end-user to view, edit and enter data,

■ *output and experimentation sections* to present model results, derived data, summaries and results pertaining to multiple cases, and

■ *control sections* for guiding the model configuration, the execution of model runs, and the printing of customized reports.

15.4 Interaction

This section

This section gives you some further design guidelines which will have a positive impact on the quality of interaction with end-users.

Know your users

One of the most important principles in user interface design is to know your users. When you consider that the interface is an intermediary between your model and the end-user, you will realize that it is a means of communication. Therefore it is essential that you carefully

■ identify the needs of your users,

■ study their standards of communication,

■ consider their level of knowledge of the application area, and

■ recognize their abilities with their computer.

The more you can accommodate your end-users' needs, the more it will reduce their learning time and improve their acceptance of the system.

Once you know your users, you will know how to address them in the interface. Several relevant aspects are: *What to emphasize*

- the symbols or text used on buttons to indicate their actions,
- the amount of guidance in the form of message dialog boxes,
- the existence of fixed sequences to carry out certain tasks,
- the existence and style of feedback messages, and
- the use of existing color conventions or symbols for certain products or status parameters.

The initial interaction with your end-users in an interface should occur without any knowledge on their part. That is why you should create a start-up procedure that runs automatically on opening the project. Typical actions that may be included in such a procedure are: *Start-up procedure*

- importing relevant data,
- executing required initial calculations,
- opening the correct first page, and
- setting up the first dialog box.

Users can become frustrated and discouraged if they work with a system in which solutions can become infeasible, input errors are not detected, or results somehow get lost. You could improve your interaction with the user by applying the following guidelines. *Interaction should be robust*

- Declare upper and lower bounds for parameters and variables. When your users enter values outside these bounds, Aimms will automatically produce an error message.
- Write error checks in procedures. These procedures can be called after a user updates a data object. If an error is detected, the procedure can issue a message.
- Provide clear and explicit diagnostics once an error is detected.
- If your end-users are not allowed to modify particular parameter values, make these parameters read-only in the interface.
- Avoid the possibility that models become infeasible by introducing extra "slack" or "surplus" variables. In addition, provide on-screen messages when these variables become positive.
- Always ask for confirmation if a user attempts to carry out an irreversible action.
- Use message dialog boxes to motivate your users to save a case after solving the model, so that they can always revert to a saved case.

Select the right object

For each identifier on a page you must select the appropriate object for its display. The following hints may be helpful.

- Use tables or scalar objects when it is important to display exact figures, or when the object will often be used for data entry.
- Use composite table objects for identifiers with many dimensions and few nonzeros. You can also use them for the display of multidimensional (sub)sets.
- Use pivot table objects if you want to view or compare identifiers with a common index domain, or if you want an object that is highly customizable by the end-user.
- Use bar charts, curves and Gantt charts for compact overviews of model results. Both curves and Gantt charts are ideal for presenting time-dependent data. Bar charts are appropriate for displaying the relative size of items. In addition to charts consider supplying extra pages with the same information in tabular form, so that exact values can be read and modified.
- Use Gantt charts when you want to combine a lot of information in one chart, or when you want to display ordering information in a sequence.
- Use network flow objects to provide a visual overview of results in any application in which flows between objects play a role.
- Use a stacked bar chart to show how a series of components add up to form a whole.
- Use a selection object or a table with 0/1 values for yes/no decisions. Both have the advantage that you can change values by single or double mouse clicks.
- Use selection objects for all situations where it is more meaningful for a user to select a description from a list rather than entering a number. If you do this, you may have to declare some extra sets, or parameters, for display purposes.
- Use AIMMS' capabilities of linking indices to element parameters to show multidimensional data. This gives you the opportunity of displaying large amounts of data in a concise way by using identifier slices fixed to one or more element parameters, and showing the data for the remaining indices only.

In all cases, it is important that you structure the information within an object in a meaningful manner. You should make deliberate decisions regarding the selection of row and column labels in a table, the choice of the x-axis in a bar chart, the number and the display of grid lines, the benefit of removing zeros, etc.

15.5 Illustrative example pages

In Figure 15.2 you find an example of a page displaying a schematic representation of the scope of the model. This not only provides the user with an insight into the process being modeled, it also serves as a menu page. The user can click on a tank or an arrow to jump to the corresponding section of pages.

A flow chart page

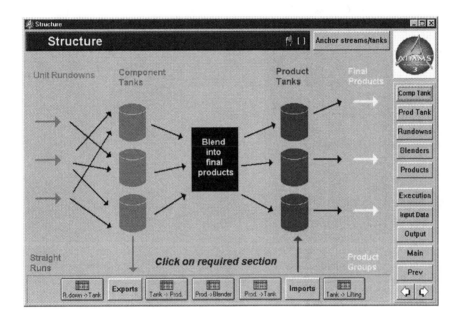

Figure 15.2: A flow chart page

A Gantt chart
page

Figure 15.3 is an example of a page displaying tasks scheduled by the model in the form of two related Gantt charts. The example page is taken from a highly interactive application that is used to quickly propagate the consequences of manual changes that are made by schedulers.

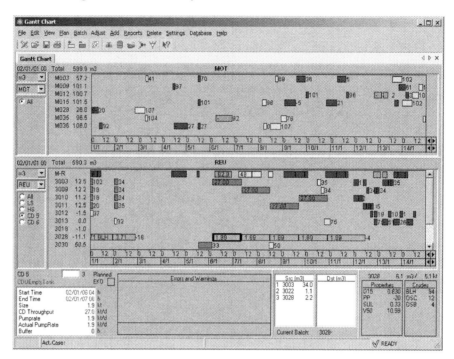

Figure 15.3: A Gantt chart page

The area on the left of each Gantt chart contains some controls that allows the user to change what is being displayed: one or more radio buttons that select the object for which tasks are being displayed and a drop-down list that allows the user to change the display unit of the numbers that are being displayed. The *x*-axis of the Gantt chart is a time axis representing both hours and days. By clicking on one of the batches in either Gantt chart, detailed information about the task is displayed in the lower part of the window. Note that the page is equipped with a user menu (and toolbar) that allows the user to perform all kinds of batch related operations.

On the page shown in Figure 15.4 is a list object that is used to display a list *A list page*
of values. The first two columns are associated with different cases, the third
column displays the difference between the two. The units associated with the
identifiers are shown to the right. In this page button areas are positioned
along the bottom and right side of the page.

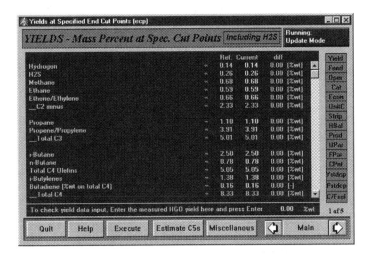

Figure 15.4: A page with lists of values and buttons

The page in Figure 15.5 contains a number of selection objects. With these ob- *A page*
jects an end-user can indicate which sections should be included in a printed *controlling a*
report. The file name object in the top right corner displays the report's file- *report*
name. The user can change this name after clicking on the icon.

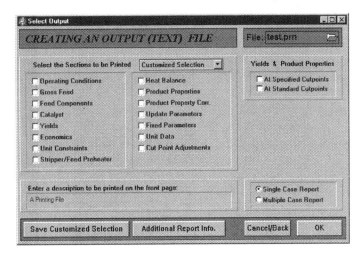

Figure 15.5: A page controlling the sections to be printed in a report

*A page
visualizing a
production flow*
Figure 15.6 shows a page from a production planning application for a steel industry reduction area (including a sintering plant and blast furnaces). The application involves the integrated planning of raw material preparation, sintering, and operations at blast furnaces. An optimization model, using a customized heuristic that sequentially solves basic MILP (mixed integer linear programming) and NLP (nonlinear programming) sub-problems, determines the quantities of raw material inputs for the production of sinter and pig iron. Nonlinear constraints are included in the model in order to compute the fuel mix for the sintering plant and blast furnaces, as well as the distribution of sinter among the blast furnaces. Integer variables were included in the model in order to represent whole train cargoes of mineral ore and operational alternatives of coal mixtures for pulverized injection.

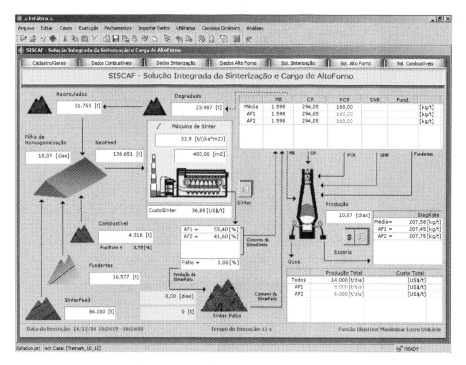

Figure 15.6: A page with several data objects, buttons and images

A collection of tables and scalar object combined with illustrative images visualizes a recommended plan for the production of sinter and pig iron out of raw materials. The seven text buttons on top of the page title can be used to navigate through the application. Note that the interface has been developed using a Brazilian character set.

Part IV

Data Management

Chapter 16

Case Management

Case management is an important part of any successful decision support application. The ability to save and work with many cases, and perform a what-if analysis by computing and analyzing the results for a large number of scenarios helps in taking the right decisions in real-life situations. This chapter introduces AIMMS' capabilities with respect to creating and managing a large database of cases, as well as its capabilities of working with data from multiple cases both within the language and in graphical data objects on end-user pages.

This chapter

16.1 What are cases?

A case forms a *complete* set of data that will enable you to restart a particular modeling application. In the simplest form of case management, which will be the subject of this chapter, a case contains the data associated with *all* identifiers in your model.

What are cases?

For more advanced use, AIMMS also supports the concepts of

Advanced use

- *case types*, with which you can instruct AIMMS to only save the data of a subset of the identifiers in your model, and
- *data categories* and *datasets*, which allow you to store common data, that is shared by multiple cases, at a single location.

Both of these subjects are discussed in Chapter 17.

In this chapter, you will find only a single case type, *All Identifiers*, which results in cases containing the data of all the identifiers in your model. This case type is automatically added to every new AIMMS project, and all dialog boxes referring to case types will default to it.

All Identifiers

When you want to work with cases in AIMMS, there are two main tools through which you can accomplish most tasks. These are:

Case management tasks

- the **Data** menu, through which you can accomplish simple case management tasks such as loading and saving cases from within your modeling application, and

■ the **Data Manager**, which you can use to manage the complete collection of cases in your application, to create batch runs of cases, or to create a selection of cases for simultaneous display.

The following two sections will discuss both tools in more detail.

16.2   Working with cases

Loading and saving

While the **Data Manager** is mainly intended for organizing and managing a large collection of datasets and cases, common end-user case management tasks such as loading case data into your model, and saving the current data in your model into cases are usually performed using the **Data** menu only. By default, the **Data** menu is available on all end-user pages.

The active case

In AIMMS, all the data that you are currently working with is referred to as the *active* case. If you have not yet loaded or saved a case, the active case is *unnamed*, otherwise the active case is *named* after the name of the last loaded or saved case on disk. If the active case is named, its name is displayed in the status bar at the bottom of the AIMMS window.

Saving a case

When you save a named active case, AIMMS will save it to the associated case on disk by default (thus overwriting its previous contents). If the active case is unnamed, or when you try to save a case using the **Data-Save Case As** menu, AIMMS will open the **Save Case** dialog box illustrated in Figure 16.1. In the

Figure 16.1: The **Save Case** dialog box

Save Case dialog box you can enter the name of the case, and, optionally, select the folder in the case tree (explained below) in which the case is to be stored. After successfully saving a case through the **Save Case** dialog box, the active case will become named.

AIMMS supports three modes for loading the data of a case:

Loading a case

- load as active,
- load into active, and
- merge into active.

These three modes of loading a case differ in

- whether they change the name of the active case or are only importing data into your current active case, and
- whether existing data is replaced by, or merged with, the loaded data.

The most frequently used mode for loading a case is loading the case *as active*, through the **Data-Load Case-As Active** menu. Loading a case as active completely replaces the active case data of all identifiers in the loaded case with their stored values. Data of identifiers that are not stored in the case, remain unchanged. In addition, the active case will be named after the loaded case. Before loading a case as active, AIMMS will ask you whether the current active case data needs to be saved whenever this is necessary.

Load as active

Loading a case *into active* (through the **Data-Load Case-Into Active** menu) is completely identical to loading a case as active, with the exception that the name of the active case will not be changed. Thus, by loading data into the active case you can replace part, or all, of the contents of the active case with data obtained from another case.

Load into active

Merging a case *into active* (through the **Data-Load Case-Merge Into Active** menu) does not change the name of the active case either. Merging a case into active partially replaces the data in the active case with only the nondefault values stored in the loaded case. Data in the active case, for which no associated nondefault values exist in the loaded case, remain unchanged.

Merge into active

Using the **Data-New Case** menu item, you can instruct AIMMS to start a new, unnamed, active case. However, the data in the active case will remain *unchanged*. If you also want to remove all data from the active case, you can accomplish this from within your model using the EMPTY statement. Before starting a new case, AIMMS will ask you whether the current active case data needs to be saved.

Starting a new case

16.3   Managing cases with the Data Manager

*The Data
Manager*

The management of all cases (and also datasets) stored within a particular
AIMMS project is done through the AIMMS **Data Manager**. It offers you a tree-
based view of all the cases created in your application, as illustrated in Figure 16.2. The figure displays a subset of the entire collection of cases. The
node name of the current active case is displayed in bold.

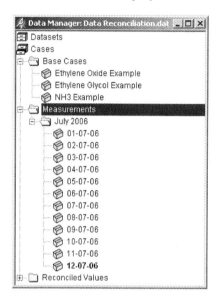

Figure 16.2: The AIMMS **Data Manager**

Managing cases

Below the case tree in the **Data Manager**, you can add any hierarchical substructure of folders to organize the available cases as you see fit. This allows
you, for example, to subdivide the entire collection of cases by user or by any
other criteria that you, or your end-users, deem relevant. Within this hierarchical structure of folders you can insert, copy and delete cases, or move
cases around in the same manner as in any of the other AIMMS trees (see also
Section 4.3).

*Creating cases
in the Data
Manager*

In addition to creating cases through the **Data** menu, you can also create new
cases in the **Data Manager** itself. Creating cases in this manner can simply be
accomplished by adding a node to the case tree, in a similar fashion as with
any other tree. When you create a case in the **Data Manager**, it will not contain
any data initially.

Creating cases in the **Data Manager** is particularly useful when you want to quickly compose a large number of scenarios which all make use of a number of datasets containing common data shared by all scenarios. Including datasets in a case created in the **Data Manager** is discussed in full detail in Section 17.3.

Including datasets

Within the **Data Manager**, you can view or modify the properties of all existing cases through the case **Properties** dialog box illustrated in Figure 16.3. This

Viewing the case properties

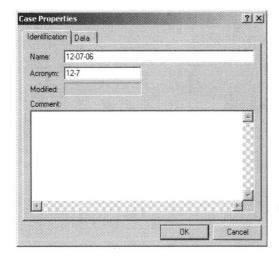

Figure 16.3: The case **Properties** dialog box

dialog box lets you provide further identification information for the case, such as an acronym for use in a multiple-case display (see Section 16.5.1) and an explanatory comment. In addition, in the **Data** tab, you can view or modify the case type (see Section 17.1) associated with the case, as well as view the datasets included in the case (see also Section 17.2.1).

16.4 Executing batch runs

The case facilities in AIMMS make it very easy to perform a what-if analysis on your model. When you store the input data for each scenario in a separate case, performing a what-if analysis boils down to running your model for a batch of cases, and comparing the results stored in these cases.

What-if analysis

After you have created a number of cases containing the (input) scenarios for a what-if analysis, you need to compute the optimal solution for all of these scenarios. If you have only a few scenarios which solve relatively quickly, you can perform this interactively through the following sequence of actions:

Interactive batch

■ load the case data,

- execute the procedure that computes the model results, and
- save the results back to the case.

Batch runs

If there are many scenarios to be solved, or if the solution time of each individual scenario is long, AIMMS offers facilities to perform the actual what-if analysis by creating and executing a *batch run* of scenarios. Such a batch run can be executed over night, or at any other time that you do not need your computer.

Adding cases to a batch run

You can add one or more selected cases from within the **Data Manager** to a batch run through the **Edit-Add To-Batch** menu in the **Data Manager**. This will open the **Batch Run** window illustrated in Figure 16.4. This window shows the

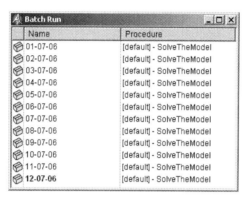

Figure 16.4: The **Batch Run** window

cases that have been already added to the batch run, along with the procedures within the model that must be executed for each case in the batch run to compute the model results.

Managing a batch run

Within the **Batch Run** window, you can further manipulate the current list of cases. You can modify the order of execution by changing the case order in the list of cases, delete cases from the list, or add cases to the batch run by dragging them from within the **Data Manager** into the **Batch Run** window.

Assigning a procedure

In addition, the **Batch Run** window lets you specify a specific procedure for each case in the list that has to be executed to compute the model results for that particular case in the batch run. You can modify the batch procedure for a case through the **Run-Procedure** menu, which will let you select a procedure from the list of all available procedures in the model.

If you do not specify a specific procedure to be executed for a particular case, AIMMS will execute a default procedure. As the developer of a project, you can specify this default procedure through the **Run-Default Procedure** menu, and AIMMS stores this information along with the project. End-users are not allowed to modify the default batch procedure. In the **Batch Run** window, AIMMS displays the default batch procedure for every case for which you have not selected a case-specific batch procedure.

The default batch procedure

After you have composed a batch of cases to your satisfaction, you can start the execution of the batch through the **Run-Start Batch** menu. This will open the dialog box shown in Figure 16.5. It displays the total number of cases in

Starting a batch

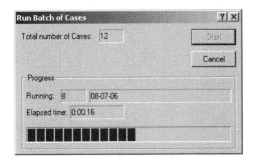

Figure 16.5: The **Run Batch** dialog box

the batch, some information about the case currently being executed, as well as the overall progress of the batch run. By pressing the **Cancel** button in the dialog box, you can interrupt the execution of a running batch.

During a batch run, AIMMS will perform the following actions for every case in the batch run:

Executing a single case

- load the case,
- run the specified or default batch procedure, and
- save the case in order to store the model results.

16.5 Managing multiple case selections

After you have executed a batch run (or when you have created several cases manually), AIMMS allows you to simultaneously view the results of several cases within the graphical user interface. In addition, it is possible to reference data from multiple cases within the modeling language, enabling you to perform advanced forms of case comparison.

Viewing batch results

Multiple case selections

AIMMS offers a special window, similar to the **Batch Run** window, to construct a selection of cases to which you want simultaneous access either from within the graphical user interface or from within the model itself. You can add one or more selected cases from within the **Data Manager** to the multiple case selection through the **Edit-Add To-Multiple Cases** menu in the **Data Manager**. This will open the **Multiple Cases** window illustrated in Figure 16.6. It shows

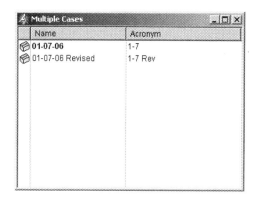

Figure 16.6: The **Multiple Cases** window

the current contents of the multiple case selection. As with the **Batch Run** window, you can modify the order of the displayed cases, and add or delete cases from the collection.

The case acronym

The acronym column in the **Multiple Cases** window displays the acronym associated with each case in the selection. You can assign an acronym to an AIMMS case through its **Properties** dialog box (see Section 16.2). Whenever available, AIMMS will use the acronym in multiple case displays within the graphical user interface.

16.5.1   Viewing multiple case data

Viewing multiple case data

The prime use of multiple case selection takes advantage of AIMMS' capability of displaying data from multiple cases within its graphical objects. Figure 16.7 illustrates a table which displays the contents of a single identifier for all the cases in the case selection shown in Figure 16.6.

Creating multiple case objects

You can turn a data object, in the graphical end-user interface, into a multiple case object by checking the multiple case property in the object-specific options in the object **Properties** dialog box. Figure 16.8 illustrates the object-specific **Properties** dialog box of a table object. As a result of enabling multiple case display, the object will be extended with one additional virtual dimension, the case index, which will be displayed in a standard way.

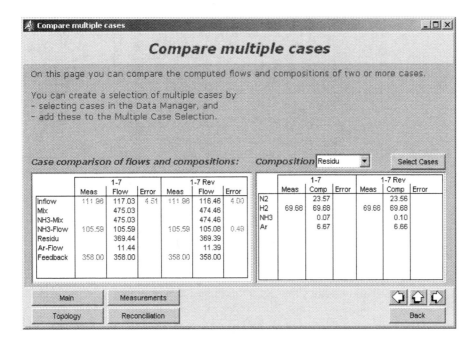

Figure 16.7: Example of a multiple case object

AIMMS only supports the display of multiple case data in object types for which *Restrictions* the added dimension can be made visible in a well-defined manner. The most important object types that support multiple case displays are tables, pivot tables, curves, bar charts and scalar objects. Because of the extra dimension, the bar chart object is only able to display multiple case data for scalar and 1-dimensional identifiers. During a single case display, a bar chart can also be used to view 2-dimensional identifiers.

16.5.2 Case referencing in the language

In addition to viewing data from multiple cases as graphical objects in the *Using inactive* graphical user interface, AIMMS also allows you to reference the data of cases *case data* that are not currently active within the model. This allows you, for instance, to perform advanced forms of case differencing by comparing the current values of particular identifiers in your model with the corresponding values stored in an inactive case.

The collection of all cases available in the **Data Manager**, is available in the *The set AllCases* AIMMS language through the predefined integer set AllCases. Each case in the **Data Manager** is represented by an integer element in this set, and, as explained in the Section 18.4.5, AIMMS offers several built-in functions to obtain additional information about a case through its case number.

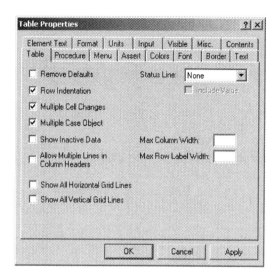

Figure 16.8: Table-specific **Properties** dialog box

Referencing
case data

You can reference the values of specific identifiers within a particular case by simply prefixing the identifier name with an index or element parameter in the set AllCases. Thus, if cs is an index in the set AllCases, the following simple assignment will inspect every case, and store the values of the variable Transport(i,j) stored in that case in the parameter CaseTransport, which has one additional dimension over the set of AllCases.

```
CaseTransport(cs,i,j) := cs.Transport(i,j);
```

Advanced case
comparison

The capability of referencing inactive case data, enables you to perform advanced forms of case comparison, which would be hard to accomplish without the AIMMS facilities for case referencing. As an example, consider the following statement.

```
RelativeDiff(cs,i,j) := (cs.Transport(i,j) - Transport(i,j)) /$ Transport(i,j);
```

It computes the relative difference between the current values of the variable Transport(i,j) and those values stored for each case on disk. You can display this data, for instance, in the graphical user interface.

The set Current-
CaseSelection

AIMMS stores the case selection constructed in the **Multiple Case Selection** dialog box discussed in the previous section in the predefined set CurrentCase-Selection, which is a subset of the set AllCases. Thus, you can very easily apply the above possibilities to only the cases selected by your end-users in the **Multiple Cases** window. The following statement illustrates a small adaptation of the previous example to restrict the computation of the relative difference to only the cases in CurrentCaseSelection.

```
RelativeDiff(cs in CurrentCaseSelection,i,j) :=
    (cs.Transport(i,j) - Transport(i,j)) /$ Transport(i,j);
```

Chapter 17

Advanced Data Management

The previous chapter discussed AIMMS' basic case management capabilities. These capabilities are sufficient for most projects. Understanding the advanced concepts introduced in this chapter will require a considerable time investment on your part. These concepts are only required when data efficiency and data security play a crucial role in your application.

Time investment required

This chapter introduces the more advanced concepts of *case types*, *data categories* and *datasets*. These concepts provide you with a flexible framework to create cases containing only a subset of the data in your model, to create data snapshots representing a particular functional aspect within a model, or to store common data, that is shared by multiple cases, at a single location. In addition, the chapter discusses advanced issues such as case security, AIMMS' facilities to import and export case data or to refer to the collection of cases and dataset from within the modeling language, and whether to use AIMMS cases or store data in a commercial database.

This chapter

17.1 Case types

A case type defines a *subset* of model identifiers which

What are case types?

- are to be stored in a case, and
- are sufficient to restore the global state of the application for a particular purpose.

You can use case types to decrease the (physical) size of cases on disk, when only a subset of all data is sufficient to restore the state of your model-based application. In addition, when the successful execution of your model consists of several phases (for data entry or for computing a (partial) solution), different case types can be used to create cases that contain only the data necessary to restart a particular phase of the modeling application.

*Data
Management
Setup*

The **Data Management Setup** window is the developer tool for specifying the collection of available case types which you think are relevant for your modeling application. As illustrated in Figure 17.1, the **Data Management Setup** window lets you manage a single list of case types. You can open it through

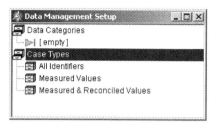

Figure 17.1: The **Data Management Setup** window

the **Tools-Data Management Setup** menu, or through the **Data Management Setup** button on the project toolbar.

*Creating case
types*

Below the *Case Types* node in the **Data Management Setup** window you can add new case type nodes to the list of already available case types. With each case type, you can associate a collection of model identifiers. With any new project AIMMS will automatically create the case type *All Identifiers*, which represents the complete collection of model identifiers currently present in the model.

*Adding model
identifiers*

To associate model identifiers with a case type you can add a list of individual identifiers to the case type either

- by dragging the identifiers from within the model explorer onto the case type node, or
- by modifying the properties of the case type.

After you have added identifiers to a case type, you can always view or edit its current content in the case type **Properties** dialog box, as illustrated in Figure 17.2. When you select a (named declaration) section in this dialog box, you get the choice to implicitly or explicitly add all identifiers to the list that are contained in this section.

*Case types and
libraries*

Cases in AIMMS represent a *global* state of your application for a particular purpose. AIMMS, therefore, only allows you to define case types for the main project. As a consequence, when adding identifiers to a case type, you can only add identifiers to the case type that

- are declared in the main project, or
- are part of the interface of a library included in the AIMMS project.

If your application requires that a case type also contains identifiers that are private to a library, you can accomplish this by defining a data category for

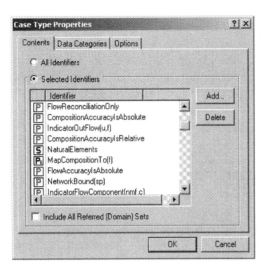

Figure 17.2: The case type **Properties** dialog box

that library containing the relevant private identifiers. Subsequently, you can add that data category to the case type, as discussed in Section 17.2.1.

When you save a case through the **Data-Save Case** or **Data-Save Case As** menus in the presence of multiple case types, the **Save Case** dialog box (illustrated in Figure 17.3) requires that you specify a case type in addition to the name of the case to be created. Similarly, the **Load Case** dialog box enables you to select from cases of all case types, or to filter on cases of a specific case type.

Selecting the case type

Figure 17.3: Saving a case with a specific case type

Preset the case type

If you do not want your end-users to select a case type themselves either when saving or loading a case, you can preset the case type from within the AIMMS modeling language through the predefined element parameter CurrentDefault-CaseType. When this element parameter has a value, AIMMS will remove thecase type drop-down list in the **Save Case** and **Load Case** dialog boxes, and use the case type specified through CurrentDefaultCaseType instead. For complete details on the parameter CurrentDefaultCaseType refer to Section 18.4.5.

17.2   Data categories and datasets

What are data categories?

A data category is a *subset* of model identifiers associated with a particular *functional aspect* of an application. For instance, you can create data categories that hold all identifiers defining the problem topology, or that define a supply and demand scenario within your application.

What are datasets?

A dataset is a *data instance* associated with a particular data category, similarly as a case is an instance of a case type. AIMMS lets you maintain multiple datasets with a data category, each dataset representing a particular version of the data. Thus, you can create datasets that define the problem topology for different regions, or that hold different scenarios for supply and demand.

Compare to case types and cases

Data categories and datasets are in many aspects similar to case types and cases. The major difference is that a

  ■ case type can include one or more data categories, and, as a consequence,
  ■ cases can be built up from multiple datasets.

Section 17.2.1 discusses in detail how you can exploit this feature.

Creating data categories

Similarly to case types, you must specify the data categories used in your model through the **Data Management Setup** tool (see also Section 17.1). Below the *Data Categories* node in the **Data Management Setup** window you can add a list of all the data category nodes that are necessary for your modeling application. In addition, AIMMS allows you to create data categories for all library projects included in your application. Figure 17.4 illustrates the list of data categories for the example application used throughout this User's Guide.

Adding identifiers

You must associate a list of model identifiers with every data category, which together completely represent the particular functional aspect of the model expressed by that data category. There are two methods for associating model identifiers with a particular data category:

  ■ you can simply drag a selection of identifiers from the **Model Explorer** (possibly obtained through the identifier selection tool) to the data category node, or

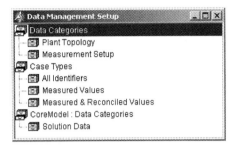

Figure 17.4: Data categories in the **Data Management Setup** tool

■ you can open the **Properties** dialog box of the data category, and modify its contents in the **Contents** tab, as illustrated in Figure 17.5.

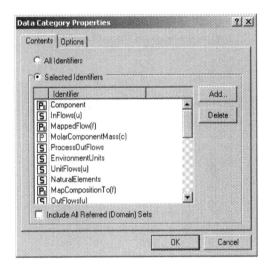

Figure 17.5: Data category **Properties** dialog box

After you have added identifiers to a data category, you can always view its current contents through the **Properties** dialog box of that data category.

For data categories associated with the main project, i.e. those listed under the *Data Categories* node in the **Data Management Setup** window, AIMMS only allows you to add identifiers that

Global data categories and libraries

■ are declared in the main project, or
■ are part of the interface of a library included in the AIMMS project.

This restriction ensures that a developer of a library project can change the internal implementation of the library without disrupting the case functionality of the main application.

Data categories
in libraries

If a data category is defined within a library project, you can add every identifier declared in the library to it. By including such a data category into a
case type, as described in Section 17.2.1, you can ensure that the entire internal state of library, that is necessary to continue working with the functional
aspect expressed by the data category, is saved in a case.

Reflected in
Data Manager

Whenever you have specified one or more data categories in the **Data Management Setup** tool, AIMMS will automatically add a *Datasets* root node to the
Data Manager. Directly below the *Datasets* node, AIMMS will add a node for
each data category that you have created with the **Data Management Setup**
tool, whether defined in the main project or in any of its included library
projects. If you have not (yet) created data categories, the *Datasets* node will
not be present in the **Data Manager** at all. Figure 17.6 illustrates an example
of the **Data Manager** associated with the data management configuration of
Figure 17.4.

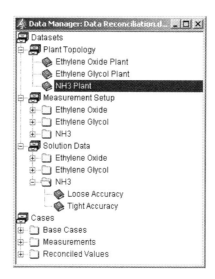

Figure 17.6: The AIMMS **Data Manager**

Managing
datasets

Below each of the data category nodes in the **Data Manager**, AIMMS allows you
to add one or more datasets associated with that data category. In addition,
AIMMS allows you to add additional folders below these nodes to provide further structure to a collection of datasets, as illustrated in Figure 17.6. However,
unlike the collection of cases (which can be structured regardless of their case
type) datasets are always strictly separated by data category in the **Data Manager**. Thus, you cannot move or copy datasets from one data category node to
another, as these represent an entirely different subset of identifiers.

17.2.1 Using data categories and datasets in cases

In addition to constructing a case type from model identifiers only (see Section 17.1), AIMMS also allows you to build up a case type from one or more data categories, or even to combine both ways of constructing a case type. You can add a data category to a case type through the case type **Properties** dialog box, as illustrated in Figure 17.7. The dialog box will let you choose both

Case types and data categories

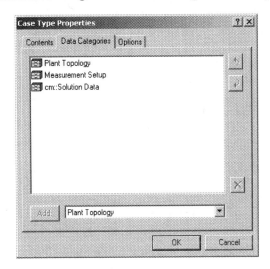

Figure 17.7: Case type **Properties** dialog box

from data categories defined in the main project and those defined in library projects.

With data categories in a case type, the complete collection of identifiers that is actually stored inside a case, consists of the union of

Case type identifiers

- the list of individual identifiers added to the case type, and
- the lists of identifiers associated with all data categories that have been added to the case type.

By default, AIMMS will store the data for all identifiers included in the case type in every case of that type.

However, when a case type contains one or more data categories, AIMMS also allows you to create *compound* cases that include references to associated datasets. As a result, AIMMS will no longer store the corresponding identifier values in the case itself, but in referenced datasets instead. Thus, dataset referencing allows the data stored in such a dataset to be shared by multiple cases.

Sharing data in cases

Sharing is space efficient   Both cases and datasets are stored on disk in a data manager file (see Section 17.6). The size of this file can be drastically reduced if you store shared data in datasets wherever possible, and include references to these datasets in your cases. As an example, if you have m topology datasets and n compatible supply and demand datasets, you can easily combine these datasets to create (input data for) $m \times n$ cases at almost no additional storage cost.

Changes are shared   In addition to storage efficiency, the use of shared datasets between AIMMS cases will ensure that a change to an identifier in a shared dataset in a single case is automatically propagated to all other cases that include the same dataset as well. This prevents you from having to go through all the individual cases to repeatedly make the identical change.

17.3 Working with datasets

Active datasets   In AIMMS, all data associated with the identifiers contained in a data category are referred to as the *active* dataset for that data category. If you have not yet loaded or saved a dataset for a particular data category, the associated active dataset is *unnamed*, otherwise the active dataset is *named* (after the name of the last loaded or saved dataset on disk).

Saving a dataset   When you save a named active dataset, AIMMS will save it to the associated dataset on disk by default (thus overwriting its previous contents). If an active dataset is unnamed, or when you try to save a dataset using the **Data-Save Dataset As** menu, AIMMS will open the **Save Dataset** dialog box illustrated in Figure 17.8. In the **Save Dataset** dialog box you must select the data category

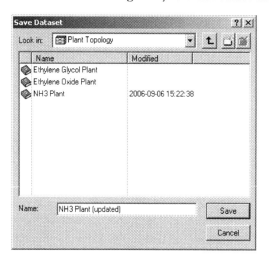

Figure 17.8: The **Save Dataset** dialog box

to be saved and enter the name of the associated dataset. Optionally, you can select a folder below the data category in which the dataset is to be stored. After successfully saving a dataset through the **Save Dataset** dialog box, the active dataset will become named.

Through the **Data-Load Dataset** menu, AIMMS allows you to load the data as-   *Loading a*
sociated with individual datasets into your model. As with loading the data of   *dataset*
a case (see Section 16.2), AIMMS supports three modes of loading a dataset:

■ as active,
■ into active, and
■ merge into active.

The actions of these modes of loading datasets are the same as for loading cases. As with cases, AIMMS will ask, before loading another dataset as active, whether the data in the current active dataset must be saved.

Within the **Data Manager** you can get an overview of the current (named) active   *View active case*
case and datasets. As illustrated in Figure 17.9, AIMMS will display the current   *and datasets*
named active case and datasets in bold typeface.

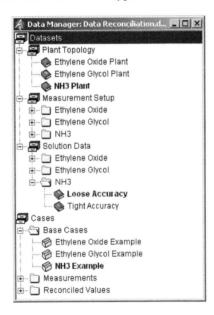

Figure 17.9: Active case and datasets

17.3.1   Datasets in compound cases

*Compound
cases*

Whenever a case type is (partially) composed of one or more data categories, loading an associated compound case will affect the corresponding (named) active datasets, whereas saving a compound case will be affected by the current active datasets. This section will explain the fine details.

*Saving a
compound case*

When you save a compound case, the active dataset for each included data category determines whether the corresponding data is saved on disk in a dataset or in the compound case itself. The following rules apply for each data category included in the case type.

- If an active dataset is named, the corresponding data is saved on disk in the named dataset, and a reference to the named dataset is stored in the compound case.
- If an active dataset is unnamed, the corresponding data is saved in the compound case itself.

Case save as

If you use the **Data-Save Case As** menu to save a compound case, the **Save Case** dialog box only lets you specify the name of the compound case itself. If you want to store data category data in named datasets, and refer to these in the compound case, you must explicitly save these datasets through the **Data-Save Dataset As** menu before saving the case.

*Loading a
compound case*

When you load a compound case, the included data or dataset references in the compound case affect the active datasets after loading. The following rules apply for each data category included in the case type.

- If the compound case contains a reference to a dataset for the data category, then the data from that dataset is loaded, and the active dataset is named after that dataset.
- If the compound case itself contains the data for the data category, then the associated data is loaded from the case, and the active dataset becomes unnamed.

*Overlapping
data*

Whenever the case type and/or the data categories in that case type refer to the same identifiers, you should be aware that the order of loading is as follows.

- First, the data stored in the compound case is loaded.
- Hereafter, the data of the included datasets are loaded using the order of the data categories as they appear in the **Data Management Setup** tool.

For all data categories not included in the compound case type, the associated active datasets are not changed by loading the compound case. You should note, however, that part of their data may be overwritten when the sets of identifiers associated with both a data category and the compound case type are overlapping.

Remaining datasets

In the **Data Manager**, you can modify the dataset references stored in a compound case, using the **Data** tab in the **Properties** dialog box of the case, as illustrated in Figure 17.10. Changing these dataset references is only useful if

Modifying dataset references

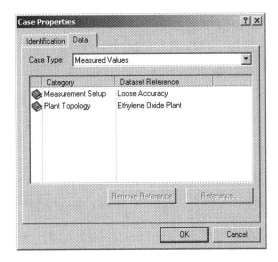

Figure 17.10: The **Data** tab in the case **Properties** dialog box

you intend to subsequently load the case using the newly specified datasets. *It has no effect on saving the compound case, since* AIMMS *will always overwrite the dataset references in a compound case according to the current active datasets*, as described above.

Modifying the dataset references manually through the **Properties** dialog box is particularly useful in combination with the possibility to create new cases by simply inserting new case nodes to the **Data Manager** tree. Combining both features allows you to quickly compose a large number of scenarios which all consist of input data taken from various combinations of existing datasets.

Composing scenarios

17.4 Advanced use of the Data Manager

In addition to viewing or modifying the contents of a data category or case type, the **Properties** dialog boxes for data categories and case types also let you optionally specify a user-defined *load procedure*. Through such a load procedure (which should only be needed for extraordinary case management

Load procedures

tasks) you can perform any further (implied) initialization statements that are necessary when a dataset or case has been loaded.

Required prototype

Each load procedure should have a single argument, an element parameter in either the predefined set AllCases or AllDatasets, referring to the case or dataset currently being loaded. The wizard used to select a load procedure in the **Properties** dialog box, will automatically restrict the choices to those procedures that match the required prototype.

New case or dataset

The load procedure is also called upon starting a new case or dataset. In that event, the argument of the load procedure refers to the empty element. You can use this feature, for instance, to empty the contents of the active case or dataset.

Datafile support

If you need further information on the specific data file passed to a load procedure, you can use the functions described in Section 18.4.5 to obtain such information about a data file. Through these functions you can get information such as the name of the data file, whether it is a case or a dataset, or any included datasets (if the data file is a case).

Included datasets

If there are load procedures for both a case type and its associated data categories, then, when a case of that type has been loaded, AIMMS will only execute the case load procedure. If you want the dataset load procedures to be executed as well, you should include the appropriate calls to these procedures in the case load procedure.

17.5 Case file security

Protecting your data

When your AIMMS-based application is used by multiple end-users, all sharing the same data management tree, read and/or write protection of the individual datasets and cases may become a relevant issue. AIMMS offers such protection by allowing you to create a database of end-users (see Section 20.3), and then letting datasets and cases be owned by individuals in this end-user database. Whenever an AIMMS application is linked to an end-user database, users must authenticate themselves before being able to use the application.

Access rights

As explained in Section 20.3, each end-user in an AIMMS end-user database must be member of a particular user group. User groups can be ordered in a hierarchical fashion. With respect to datasets and cases, AIMMS allows you to assign different access rights to

- the owner of the dataset or case,
- members of the group associated with the dataset or case,

- members of groups that lie hierarchically above or below the user group associated with the dataset or case, and
- all other users.

By default, any dataset or case will be owned by the user ID and group of the user who created it. In addition, the access rights associated with such a dataset or case will be the default access rights of the end-user (or group of end-users). These default access rights are assigned by the local user administrator in the end-user database (see also Section 20.4).

Default access rights

When you are the owner of a dataset or case it is possible to modify previously assigned access rights to a case. You can perform this task through the **Access** tab of the **Properties** dialog box of the dataset or case in the **Data Manager**, which will only be present if an end-user database is linked to your application. In the **Access** tab, displayed in Figure 17.11, you can modify the associated

Modifying access rights

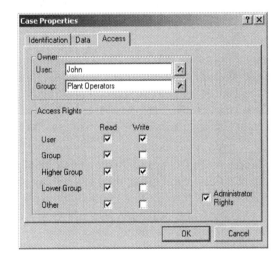

Figure 17.11: Access rights of a dataset or case

user ID and group that own the data file, as well as the access rights for each of the access categories listed above.

Normally, AIMMS will only allow you to modify the access rights of the datasets and cases that you yourself own. You can override this by checking the **Administrator Rights** check box displayed in Figure 17.11. This will pop up a password dialog box requesting the administrator password associated with the end-user database. If successful, you can modify the access rights of any dataset or case as if you were its owner.

Administrator rights

17.6   Data manager files

Case and
dataset storage

By default, AIMMS stores all cases and datasets associated with a particular modeling project in a single *data manager file* within the project directory. AIMMS allows you, however, to create or open another data manager file. This makes it possible that the cases and datasets of all end-users of your project be stored in a common data file.

Selecting a data
manager file

You can select another data manager file by means of the **File-Open-Data File** menu. Alternatively, you can create a new data manager file through the **File-New-Data File** menu. For every project, AIMMS remembers the last data manager file opened during an AIMMS session, and will reopen with the same data manager file at the beginning of a new session. Alternatively, you can indicate the data manager file with which you want to open a project as a command line argument in AIMMS. The complete list of AIMMS' command line arguments is provided in Section 19.1.

End-user
database

When your modeling application is linked to an end-user database, any newly created data manager file will also automatically be associated with that end-user database. When you try to select another data manager file, AIMMS will only allow this if the current end-user database of the application coincides with the end-user database associated with the selected data manager file.

Importing and
exporting data
files

The **Import** and **Export** facilities in the AIMMS **Data Manager** allow you to transfer a collection of datasets and cases stored in the case tree to a new data manager file, and vice versa. The import and export facilities let you easily create a backup of your data onto a floppy, and offer you a means of sending a single scenario to an interested colleague.

Exporting data
files

Through the **Export** facility AIMMS can export the selected cases and datasets in the data manager to a new data manager file. If any of the selected cases contains references to datasets which you have not explicitly selected, such datasets will be exported as well. This ensures that any exported case will refer to exactly the same data, when imported by another user. The newly created data manager file is associated with the same end-user database as the currently open data manager file, and the exported cases and datasets have the same owner and access rights as before.

Importing data
files

Through the **Import** facility AIMMS allows you to import *all* cases and datasets within a given data manager file into the current case tree. If a certain import case or dataset already exists in the case tree in which it is imported, you have the choice of overwriting the existing entry or creating a new node. AIMMS will always create a new node if you do not have permission to overwrite an existing node in the case tree. When AIMMS creates a new node for an existing entry, the name of the existing node is prefixed with the string 'Imported', followed by a number if there are more than one imported copies. Any existing entry that is overwritten will keep its current owner and access rights, while newly created cases and datasets will have the same owner and access rights as stored in the import file.

17.7 Accessing the case tree from within the model

When your modeling application depends on the use of multiple case types and/or data categories, performing data management in a consistent manner may be a task that is too involved for a casual end-user. In such cases, AIMMS allows you to setup a custom data management system yourself. From within the AIMMS language, you have access to the contents of the AIMMS case and dataset tree, as well as to all functionality for loading, saving and creating datasets and cases available through the standard end-user menus. This enables you to shield your end-users from choices which may be too involved for them to understand, and ensure that every created case is built up in a consistent manner.

Custom data management

After you created a customized data management system, you may want to restrict the end-user capabilities in the **Data Manager** on a task-by-task basis to prevent them from making inadvertent mistakes. You can accomplish this by modifying the appropriate options in the **Properties** dialog box of a case type or data category and in the global AIMMS **Options** dialog box. For instance, AIMMS allows you to completely hide datasets within the **Data Manager** if you want your case management scheme to depend on datasets, but do not want to bother your users. Modifying global AIMMS options is explained in full detail in Section 21.1.

Restricting end-user capabilities

All data categories, datasets and cases in an application are accessible in the AIMMS language through a number of predefined sets and parameters. They are:

Predefined model identifiers

- the set AllDataCategories, containing the names of all data categories defined in the data manager setup window,
- the set AllCaseTypes, containing the names of all case types defined in the data manager setup window,
- the integer set AllDataFiles, representing all datasets and cases available with a particular project,
- the set AllDatasets, a subset of AllDataFiles, representing the collection of all datasets available in the project,
- the set AllCases, a subset of AllDataFiles, representing the set of all cases available for the project,

- the indexed element parameter `CurrentDataset` in `AllDatasets` and defined over `AllDataCategories` containing the currently active datasets,
- the scalar element parameter `CurrentCase` in `AllCases`, and
- the scalar element parameter `CurrentDefaultCaseType` in `AllCaseTypes`.

Interface functions

In addition to the collection of predefined identifiers described above, AIMMS offers a complete range of data management related interface functions which you can call from within your model to perform data management tasks. These functions are described in Section 18.4.5. They allow you to perform tasks such as:

- obtaining additional information about the data categories, datasets and cases contained in one of the predefined sets described above, as available within the **Data Manager**,
- invoking functionality from the **Data Manager**'s end-user menus,
- invoking functionality from the end-user **Data** menu.

Guide your end-users

By combining the above, it is possible to guide your end-users through the process of selecting datasets for all data categories involved in a particular case, providing additional information as you see fit. Alternatively, if a case has a known and fixed structure, AIMMS allows you to build it up within the language without any user interaction.

17.8 The case tree versus databases

Case tree as database?

The features discussed in the previous section may have lead you to believe that you can use the AIMMS case tree as a database to store data owned and shared by multiple users. While this is true in principle, there are situations where the use of a true database is preferable over using the AIMMS case tree. This section discusses the issues which you should take into consideration before making a choice.

Speed

AIMMS cases are tailored to store the contents of one or more identifiers in your model quickly and easily. Therefore, storing and retrieving data through AIMMS case files is much faster than accessing the same data from a database server. In addition, setting up a link with a database is much more involved, as you need to specify a `READ` or `WRITE` statement for every individual table in the database. The increased speed of data retrieval from and to case files may be essential to gain end-user acceptance of your application.

The focus when storing data in a database is inherently different from storing data in an AIMMS case tree. Whereas the object of a database is to store and maintain a single version of a particular table to be shared by several applications, the AIMMS case tree is specifically set up to be able to easily maintain and switch between multiple versions of the data associated with a particular set of identifiers within a single model. This AIMMS feature allows you to easily perform a what-if analysis by running your model with different scenarios stored as separate case files. In addition, AIMMS allows you to simultaneously view identifier values from multiple case files within a single graphical object.

Different focus

AIMMS will ensure that any read or write action on a data file will not interfere with another user accessing the same data file at the same time. However, since AIMMS copies the data from a data file into memory, there is no guarantee that the data in the data file remains synchronized with changes made to the case file by other users. Whenever such synchronization is essential in your application, you are advised to use database technology.

Synchronization

Although the use of cases and datasets is the fastest and most convenient way to store and retrieve internal AIMMS data, it is not suitable for storage of data which the AIMMS application shares with other applications. When your AIMMS application needs input data produced by another application, or when the results of your model are input to other programs, you are strongly advised to store the data in a database.

Data sharing

Part V

Miscellaneous

Chapter 18

User Interface Language Components

Most of the functionality in the AIMMS graphical user interface that is relevant to end-users of your modeling application can be accessed directly from within the AIMMS modeling language. This chapter discusses the functions and identifiers in AIMMS that you can use within your model

This chapter

- to influence the appearance and behavior of data shown in your end-user interface, or
- to provide (or re-define) direct interaction with the end-user interface through dialog boxes, menus and buttons.

Rather than providing a complete reference of all these functions, this chapter provides you with a global overview of the functions available per functional category. A complete function reference is made available as part of the AIMMS documentation in electronic form.

18.1 Updatability of identifiers

In many applications you, as a modeler, might need to have dynamic control over the updatability of identifiers in the graphical end-user interface of your model. AIMMS provides several ways to accomplish this.

Dynamic control required

A typical example of dynamically changing inputs and outputs is when your model is naturally divided into multiple decision phases. Think of a planning application where one phase is the preparation of input, the next phase is making an initial plan, and the final phase is making adjustments to the initial plan. In such a three-layered application, the computed output of the initial plan becomes the updatable input of the adjustment phase.

Multiple phases in your application

To change the updatability status of an identifier in the graphical interface you have two options.

Indicating input and output status

- You can indicate in the object **Properties** dialog box whether all or selected values of a particular identifier in the object are updatable or read-only.
- With the set CurrentInputs you can change the global updatability status of an identifier. That is, AIMMS will never allow updates to identifiers

that are not in the set CurrentInputs, regardless of your choice in the properties form of a graphical object.

The set
CurrentInputs

The set CurrentInputs (which is a subset of the predefined set AllUpdatable-Identifiers) ultimately determines whether a certain identifier can be treated as an input identifier for objects in an end-user interface. You can change the contents of the set CurrentInputs from within your model. By default, AIMMS initializes it to AllUpdatableIdentifiers.

The set
AllUpdatable-
Identifiers

The set AllUpdatableIdentifiers is computed by AIMMS when your model is compiled, and contains the following identifiers:

- all *sets* and *parameters* without definitions, and
- all *variables* and *arcs*.

Thus, sets and parameters which have a definition can never be made updatable from within the user interface.

18.2 Setting colors within the model

Color as
indicator

An important aspect of an end-user interface is the use of color. Color helps to visualize certain properties of the data contained in the interface. As an example, you might want to show in red all those numbers that are negative or exceed a certain threshold.

Setting colors in
the model

AIMMS provides a flexible way to specify colors for individual data elements. The color of data in every graphical object in the graphical interface can be defined through an (indexed) "color" parameter. Inside your model you can make assignments to such color parameters based on any condition.

The set
AllColors

In AIMMS, all *named* colors are contained in the predefined set AllColors. This set contains all colors predefined by AIMMS, as well as the set of logical color names defined by you for the project. Whenever you add a new logical color name to your project through the color dialog box, the contents of the set AllColors will be updated automatically.

Color
parameters

Every (indexed) element parameter with the set AllColors as its range can be used as a color parameter. You can simply associate the appropriate colors with such a parameter through either its definition or through an assignment statement.

Assume that ColorOfTransport(i,j) is a color parameter defining the color of the variable Transport(i,j) in an object in the end-user interface. The following assignment to ColorOfTransport will cause all elements of Transport(i,j) that exceed the threshold LargeTransportThreshold to appear in red.

Example

```
ColorOfTransport((i,j) | Transport(i,j) >= LargeTransportThreshold) := 'Red' ;
```

18.2.1 Creating non-persistent user colors

During the start up of an AIMMS project, the set AllColors is filled initially with the collection of persistent user colors defined through the **Tools-User Colors** dialog box (see also Section 11.4). Through the functions listed below, you can extend the set AllColors programmatically with a collection of non-persistent colors, whose lifespan is limited to a single session of a project.

Non-persistent user colors

- UserColorAdd(*colorname,red,green,blue*)
- UserColorDelete(*colorname*)
- UserColorModify(*colorname,red,green,blue*)
- UserColorGetRGB(*colorname,red,green,blue*)

The argument *colorname* must be a string or an element in the set AllColors. The arguments *red*, *green* and *blue* must be scalars between 0 and 255.

You can use the function UserColorAdd to add a non-persistent color *colorname* to the set AllColors. The RGB-value associated with the newly added user color must be specified through the arguments *red*, *green* and *blue*. The function will fail if the color already exists, either as a persistent or non-persistent color.

Adding non-persistent colors

Through the functions UserColorDelete and UserColorModify you can delete or modify the RGB-value of an existing non-persistent color. The function will fail if the color does not exist, or if the specified color is a persistent color. Persistent colors can only be modified or deleted through the **Tools- User Colors** dialog box.

Deleting and modifying colors

You can obtain the RGB-values associated with both persistent and non-persistent user colors using the function UserColorGetRGB. The function will fail if the specified color does not exist.

Retrieving RGB-values

18.3 Creating histograms

The term histogram typically refers to a picture of a number of observations. The observations are divided over equal-length intervals, and the number of observed values in each interval is counted. Each count is referred to as a frequency, and the corresponding interval is called a frequency interval. The

Histogram

picture of a number of observations is then constructed by drawing, for each frequency interval, the corresponding frequency as a bar. A histogram can thus be viewed as a bar chart of frequencies.

Histogram support

The procedures and functions discussed in this section allow you to create histograms based on a large number of trials in an experiment conducted from within your model. You can set up such an experiment by making use of random data for each trial drawn from one or more of the distributions discussed in the AIMMS Language Reference. The histogram frequencies, created through the functions and procedures discussed in this section, can be displayed graphically using the standard AIMMS bar chart object.

Histogram functions and procedures

AIMMS provides the following procedure and functions for creating and computing histograms.

- HistogramCreate(*histogram-id*[,*integer-histogram*][,*sample-buffer-size*])
- HistogramDelete(*histogram-id*)
- HistogramSetDomain(*histogram-id*,*intervals*[,*left*,*width*]
                                              [,*left-tail*][,*right-tail*])
- HistogramAddObservation(*histogram-id*,*value*)
- HistogramGetFrequencies(*histogram-id*,*frequency-parameter*)
- HistogramGetBounds(*histogram-id*,*left-bound*,*right-bound*)
- HistogramGetObservationCount(*histogram-id*)
- HistogramGetAverage(*histogram-id*)
- HistogramGetDeviation(*histogram-id*)
- HistogramGetSkewness(*histogram-id*)
- HistogramGetKurtosis(*histogram-id*)

The *histogram-id* argument assumes an integer value. The arguments *frequency-parameter*, *left-bound* and *right-bound* must be one- dimensional parameters (defined over a set of intervals declared in your model). The optional arguments *integer-histogram* (default 0), *left-tail* (default 1) and *right-tail* (default 1) must be either 0 or 1. The optional argument *sample-buffer-size* must be a positive integer, and defaults to 512.

Creating and deleting histograms

Through the procedures HistogramCreate and HistogramDelete you can create and delete the internal data structures associated with each individual histogram in your experiment. Upon success, the procedure HistogramCreate passes back a unique integer number, the *histogram-id*. This reference is required in the remaining procedures and functions to identify the histogram at hand. The observations corresponding to a histogram can be either continuous or integer-valued. AIMMS assumes continuous observations by default. Through the optional *integer-histogram* argument you can indicate that the observations corresponding to a histogram are integer-valued.

For every histogram, AIMMS will allocate a certain amount of memory for storing observations. By default, AIMMS allocates space to store samples of 512 observations at most. Using the optional *sample-buffer-size* argument, you can override the default maximum sample size. As long as the number of observations is still smaller than the sample buffer size, all observations will be stored individually. As soon as the actual number of observations exceeds the sample buffer size, AIMMS will no longer store the individual observations. Instead, all observations are then used to determine the frequencies of frequency intervals. These intervals are determined on the basis of the sample collected so far, unless you have specified interval ranges through the procedure HistogramSetDomain.

Sample buffer size

You can use the function HistogramSetDomain to define frequency intervals manually. You do so by specifying

Setting the interval domain

- the number of fixed-width *intervals*,
- the lower bound of the *left*-most interval (not including a left-tail interval) together with the (fixed) *width* of intervals to be created (optional),
- whether a *left-tail* interval must be created (optional), and
- whether a *right-tail* interval must be created (optional).

The default for the *left* argument is -INF. *Note that the* left *argument is ignored unless the* width *argument is strictly greater than 0.* Note that the selection of one or both of the tail intervals causes a corresponding increase in the number of frequency intervals to be created.

Whenever an observed value is smaller than the lower bound of the left-most fixed-width interval, AIMMS will update the frequency count of the left-tail interval. If the left-tail interval is not present, then the observed value is lost and the procedure HistogramAddObservation (to be discussed below) will have a return value of 0. Similarly, AIMMS will update the frequency count of the right-tail interval, when an observation lies beyond the right-most fixed-width interval.

Use of tail intervals

Whenever, during the course of an experiment, the number of added observations is still below the sample buffer size, you are allowed to modify the interval ranges. As soon as the number of observations exceeds the sample buffer size, AIMMS will have fixed the settings for the interval ranges, and the function HistogramSetDomain will fail. This function will also fail when previous observations cannot be placed in accordance with the specified interval ranges.

Adjusting the interval domain

You can use the procedure HistogramAddObservation to add a new observed value to a histogram. Non-integer observations for integer-valued histograms will be rounded to the nearest integer value. The procedure will fail, if the observed value cannot be placed in accordance with the specified interval ranges.

Adding observations

Obtaining
frequencies

With the procedure HistogramGetFrequencies, you can request AIMMS to fill a one-dimensional parameter (slice) in your model with the observed frequencies. The cardinality of the index domain of the frequency parameter must be at least as large as the total number of frequency intervals (including the tail interval(s) if created). The first element of the domain set is associated with the left-tail interval, if created, or else the left-most fixed-width interval.

Interval
determination

If you have provided the number of intervals through the procedure Histogram-SetDomain, AIMMS will create this number of frequency intervals plus at most two tail intervals. Without a custom-specified number of intervals, AIMMS will create 16 fixed-width intervals plus two tail intervals. If you have not provided interval ranges, AIMMS will determine these on the basis of the collected observations. As long as the sample buffer size of the histogram has not yet been reached, you are still allowed to modify the number of intervals prior to any subsequent call to the procedure HistogramGetFrequencies.

Obtaining
interval bounds

Through the procedure HistogramGetBounds you can obtain the left and right bound of each frequency interval. The bound parameters must be one-dimensional, and the cardinality of the corresponding domain set must be at least the number of intervals (including possible left- and right-tail intervals). The lower bound of a left-tail interval will be -INF, the upper bound of a right-tail interval will be INF.

Obtaining
statistical
information

The functions HistogramGetObservationCount, HistogramGetAverage, Histogram-GetDeviation, HistogramGetSkewness and HistogramGetKurtosis provide further statistical information about the sample collected so far, such as the total number of observations, the arithmetic mean of all observed values, their standard deviation, their skewness and their kurtosis coefficient.

Example

In the following example, a number of observable outputs o of a mathematical program are obtained as the result of changes in a single uniformly distributed input parameter InputRate. The interval range of every histogram is set to the interval [0,100] in 10 steps, and it is assumed that the set associated with index i has at least 12 elements.

```
for (o) do
    HistogramCreate( HistogramID(o) );
    HistogramSetDomain( HistogramID(o), intervals: 10, left: 0.0, width: 10.0 );
endfor;

while ( LoopCount <= TrialSize ) do
    InputRate := Uniform(0,1);
    solve MathematicalProgram;
    for (o) do
        HistogramAddObservation( HistogramID(o), ObservableOutput(o) );
    endfor;
endwhile;
```

```
for (o) do
    HistogramGetFrequencies( HistogramID(o), Frequencies(o,i) );
    HistogramGetBounds( HistogramID(o), LeftBound(o,i), RightBound(o,i) );
    HistogramDelete( HistogramID(o) );
endfor;
```

18.4 Interfacing with the user interface

At particular times, for instance during the execution of user-activated procedures, you may have to specify an interaction between the model and the user
through dialog boxes and pages. To accommodate such interaction, AIMMS
offers a number of *interface functions* that perform various interactive tasks
such as

*Interface
functions*

- opening and closing pages,
- printing pages,
- file selection and management,
- obtaining numeric, string-valued or element-valued data,
- selecting, loading and saving cases and datasets, and
- execution control.

All interface functions have an integer return value. For most functions the
return value is 1 (success), or 0 (failure), which allows you to specify logical
conditions based on these values. If you are not interested in the return value,
the interface functions can still be used as procedures.

Return values

There are some interface functions that also return one or more output arguments. In order to avoid possible side effects, the return values of such
functions can only be used in scalar assignments, and then they must form
the entire right hand side.

*Limited use in
certain cases*

Whenever an interface function fails, an error message will be placed in the
predefined AIMMS string parameter CurrentErrorMessage. The contents of this
identifier always refer to the message associated with the last encountered
error, i.e. AIMMS does not clear its contents. Within the execution of your
model, however, you are free to empty CurrentErrorMessage yourself.

*Obtaining the
error message*

The following statements illustrate valid examples of the use of the interface
functions FileExists, DialogAsk, and FileDelete.

Example

```
if ( FileExists( "Project.lock" ) ) then
    Answer := DialogAsk( "Project is locked. Remove lock and continue?",
                            Button1 : "Yes", Button2 : "No" ) ;

    if ( Answer = 1 ) then
        FileDelete( "Project.lock" ) ;
    else
        halt;
    endif ;
endif ;
```

The interface function DialogAsk has a return value of 1 when the first button is pressed, and 2 when the second button is pressed.

18.4.1 Page functions

Model page control

The possibility of opening pages from within a model provides flexibility compared to page tree-based navigation (see Section 12.1.2). Depending on a particular condition you can decide whether or not to open a particular page, or you can open different pages depending on the current status of your model.

Page functions

The following functions for manipulating pages are available in AIMMS.

- PageOpen(*page*)
- PageOpenSingle(*page*)
- PageClose([*page*])
- PageGetActive(*page*)
- PageGetFocus(*page*,*tag*)
- PageSetFocus(*page*,*tag*)
- PageSetCursor(*page*,*tag*,*scalar-reference*)
- PageRefreshAll
- PageGetChild(*page*, *result-page*)
- PageGetParent(*page*, *result-page*)
- PageGetPrevious(*page*, *result-page*)
- PageGetNext(*page*, *result-page*)
- PageGetTitle(*page*, *title*)
- PageGetUsedIdentifiers(*page*, *identifier-set*)

The arguments *page*, *result-page*, *tag* and *title* are string arguments. The argument *scalar-reference* is a scalar reference to a data element associated with an (indexed) identifier and the output argument *identifier-set* must be a subset of AllIdentifiers.

Opening and closing pages

With the PageOpen and PageClose functions you can open and close specific pages that are part of your model. The PageOpenSingle function will, in addition to opening a page, close all other pages that are currently open. You can use it, for instance, to return to the main menu of your application and close all data pages at the same time. If you do not provide a page name in the PageClose function, AIMMS will close the currently active page. To obtain the pagename of the currently active page, you can use the function PageGetActive.

The function PageSetFocus provides you with even more control over the manner in which a page is opened. Every object on a page can be tagged by means of a descriptive string. With the PageSetFocus function you can open a page and set the focus on a particular tagged object. If the execution of user-initiated procedures depends on a precise object on a page from which it is called, you can use the function PageGetFocus to obtain the current page name as well as the tag of the object which currently has the focus on that page.

Setting and getting the focus

With the function PageSetCursor you have maximum control during the opening of a page. Not only can you indicate the object on the page, but you can also specify where the cursor should be positioned within the object. You do this by entering a scalar reference to the particular data element associated within the object that should have the focus. For example, if the cursor is to be positioned within an object at the field associated with the value Transport('Amsterdam','Rotterdam'), then this value should be entered as the third argument in the function. This function can be convenient for guiding the end-user of your application through a number of interrelated pages or objects.

Setting the cursor

The functions PageOpen, PageOpenSingle, PageSetFocus, and PageSetCursor will return immediately for standard end-user pages. When you have specified that a page is a dialog page (see Section 11.3), the page will appear as a dialog box, and these interface functions block until the dialog box is closed by the user. Dialog pages allow you to construct your own customized dialog boxes, while still using the ordinary interface elements offered by AIMMS.

Dialog pages

With the PageRefreshAll function you can refresh the contents of all pages during the execution of a procedure. This is useful, for instance, when you want to show intermediate results during a long computation, or want to provide a graphical representation of the progress of a solver, updated at regular intervals (using the solver callback features discussed in the AIMMS Language Reference). Note that AIMMS will automatically refresh all pages after the user-initiated execution of a procedure has ended.

Refreshing page contents

With the functions PageGetChild, PageGetParent, PageGetPrevious and PageGet-Next you can obtain the first child page, the parent page, the previous and the next page relative to the position of the reference page named *page* in the page tree of your project (see also Section 12.1.2). The function PageGetNextInTree-Walk will provide you with the next page while traversing the page tree in a depth-first manner. This function includes hidden pages and ignores separators and can be used, for instance, to initialize the set of all pages that are

Obtaining navigation info

present in your application. If *page* is an empty string, the location of the result page will be relative to the currently active page in the graphical user interface.

Page titles

The function PageGetTitle can be used to obtain the title of a page in specified on the page **Properties** dialog box.

Listing used identifiers

With the function PageGetUsedIdentifiers you can create a list of identifiers that are used on a given page. This list consists of the identifiers being displayed on the page as well as identifiers that are used to specify object properties. This function can be used, for instance, (in combination with the function PageGetNextInTreeWalk) to programmatically generate a list of all identifiers that are used in the user interface of your application.

18.4.2 Print functions

Printing facilities

AIMMS provides a printing capability in the form of *print pages* (see Chapter 14). Ordinary pages and print pages are constructed in the same way. When you instruct AIMMS to print an ordinary page, the entire contents of the page, as you see it on the screen, are printed. When you print a print page, all interactive objects such buttons, list boxes, check boxes, radio buttons and drop-down lists are ignored. In addition, data objects on a print page that are too large to fit on a single sheet of paper, will be printed on multiple sheets.

Printing reports

You can instruct AIMMS to print any print page from within the model by using print interface functions. In addition, the print interface functions offer you the capability of composing, and printing, a customized report consisting of multiple print pages. For instance, you could use these facilities to create ready-to-go faxes on the basis of your latest scheduling results.

Print functions

The following functions are available for printing print pages in AIMMS.

- PrintPage(*page*[,*filename*][,*from*][,*to*])
- PrintStartReport(*title*[,*filename*])
- PrintEndReport
- PrintPageCount(*page*)

The arguments *page*, *filename* and *title* are string arguments. The optional arguments *from* and *to* are integer arguments.

With the `PrintPage` function you can print a single print page. If the page contains a data object for which the available data does not fit onto a single page, AIMMS will print the object over multiple pages in a row-wise manner. Through the optional *from* and *to* arguments, you can limit the page range which will actually be printed.

Printing a single page

With the optional *filename* argument you can indicate that the print output should be directed to the specified file, rather than directly sending it to the default printer. The *filename* argument is ignored when the `PrintPage` function is surrounded by calls to the `PrintStartReport` and `PrintEndReport` functions (see below). You can use the *filename* argument, for instance, to make a printed report available to others using a default filename.

Printing to file

When you want to compose a report consisting of several existing numbered print pages, you can use the `PrintStartReport` and `PrintEndReport` functions. Following a call to the `PrintStartReport` function, all pages to be printed by subsequent calls to the `PrintPage` function will be collected. They will be printed as soon as the `PrintEndReport` function is encountered. If you specify the optional *filename* argument, the output will be sent to the indicated file.

Printing customized reports

During the printing of a report AIMMS will number pages consecutively. The page number is available to you through the predefined identifier `CurrentPageNumber`. You can use it on print pages to show the page number. AIMMS will reset the page number to 1 for every *single* page printed, as well as at the beginning of a printed report. By making assignments to `CurrentPageNumber` inside a pair of calls to `PrintStartReport` and `PrintEndReport`, however, you can modify the page numbering within a printed report as you desire.

Page numbers

You can use the function `PrintPageCount`, when you are interested in the number of sheets required to print a particular print page prior to actually printing it. The function returns the number of sheets of paper needed to print the page given the current print settings and data contained on the page.

Counting pages

18.4.3 File functions

The interactive execution of your model may involve various forms of file manipulation. For instance, the user might indicate which names to use for particular input and output files, or in which directory they are (to be) stored.

File manipulation

File functions   The following functions are available for file manipulation in AIMMS.

- FileSelect(*filename*[,*directory*][,*extension*][,*title*])
- FileSelectNew(*filename*[,*directory*][,*extension*][,*title*])
- FileDelete(*filename*[,*delete_readonly_files*])
- FileCopy(*oldname*,*newname*[,*confirm*])
- FileMove(*oldname*,*newname*[,*confirm*])
- FileAppend(*filename*,*appendname*)
- FileExists(*filename*)
- FileView(*filename*[,*find*])
- FileEdit(*filename*[,*find*])
- FilePrint(*filename*)
- FileTime(*filename*,*filetime*)
- FileTouch(*filename*,*newtime*)

The arguments *filename, directory, oldname, newname* and *appendname* are string parameters. The arguments *directory, extension, title, filetime* and *newtime* are all string arguments. The optional argument *confirm* must be 0 (default) or 1, while *find* is a string argument. All optional arguments must be tagged with their formal argument name.

Directory functions   The following functions are available for directory manipulation.

- DirectorySelect(*directoryname*[,*directory*][,*title*])
- DirectoryCreate(*directoryname*)
- DirectoryExists(*directoryname*)
- DirectoryGetCurrent(*directoryname*)
- DirectoryDelete(*directoryname*[,*delete_readonly_files*])
- DirectoryCopy(*oldname*,*newname*[,*confirm*])
- DirectoryMove(*oldname*,*newname*[,*confirm*])

The arguments *directoryname, directory, title, oldname* and *newname* are all string parameters. The *directory* and *title* arguments are optional. The optional argument *confirm* must be 0 (default) or 1.

Selecting a file or directory   The functions FileSelect and FileSelectNew both open a standard file selection dialog box, and let you select either an existing or a new file. The function DirectorySelect lets you select an existing directory name or create a new one. If you do not specify a starting directory, the dialog box will start in the current working directory. If you specify a *relative* directory path, then the dialog box will start in the specified directory relative to the current directory. Using the optional extension control in the dialog you can filter the files to be shown in the dialog box. The optional *title* argument will appear in the dialog box title.

You should keep in mind that these functions only bring up a dialog box and register the user's selection and action. Depending on the button clicked (**OK** or **Cancel**) the function returns a value of either 1 or 0. What happens next depends on the way you write your code.

Clicking OK or Cancel

With the functions FileDelete, FileCopy, FileMove, DirectoryDelete, Directory-Copy and DirectoryMove you can delete a file or directory, or copy or move it to another file or directory. The function DirectoryCreate creates the given directory (without needing a dialog box). The functions FileExists and Directory-Exists let you verify whether the given file or directory exists in the file system. If you specify a relative pathname for a file or directory argument, AIMMS will assume that the path is relative to the current working directory. The working directory itself can be retrieved using the DirectoryGetCurrent function. Through the optional argument *delete_readonly_files* (with default 0) of the functions FileDelete and DirectoryDelete you can indicate whether you want read-only files to be deleted without further notice, or whether you want these functions to fail.

File and directory manipulations

The function FileTime will return the time at which a particular file was last saved. The resulting time is returned as a string with the form "*YYYY-MM-DD hh:mm:ss*". This can be transformed into any numeric time representation using the function StringToMoment discussed in the AIMMS Language Reference. The function FileTouch can be used to set the file time to the time as specified by the optional *newtime* argument. If omitted the modification time of the file is set to the current time.

File times

The delete, copy and move functions will accept wildcards (*) to delete, copy or move multiple files or directories. In these cases, the second argument must be a directory name to which the files can be copied or moved.

Use of wildcards

Using the FileView and FileEdit functions, you invoke the AIMMS editor to view or edit ASCII files from within the interface. In view mode it is not possible to modify the file. However, the **Cut, Copy, Find, Print** and **Save As** commands are still allowed. By specifying the optional *find* argument, AIMMS will search for the specified search string, and jump to its first occurrence in the selected file. You can use the FilePrint function to print an ASCII file from within your model. The file is sent to the default printer.

View, edit or print a file

18.4.4 Dialog box functions

During the execution of your model, it is very likely that you must communicate particular information with your user at some point in time. AIMMS supports two types of dialog boxes for user communication:

Two types of dialog boxes

- information dialog boxes, and
- data entry dialog boxes.

In addition to these standard dialog boxes available in AIMMS, it is also possible to create customized dialog boxes using dialog pages (see Section 11.3), and open these using the PageOpen function discussed in Section 18.4.1.

Information dialog boxes

The following functions are available in AIMMS for displaying information to the user.

- DialogMessage(*message*[,*title*])
- DialogError(*message*[,*title*])
- DialogAsk(*message,button1,button2*[,*button3*])
- DialogProgress(*message*[,*percentage*])
- StatusMessage(*message*)

The *message*, *title* and *button* arguments are strings. The *percentage* argument is a number between 0 and 100.

Dialog box on the screen

With the DialogMessage and DialogError functions you can display a dialog box containing your own message and an **OK** button. With the optional *title* argument, you can specify the title of the dialog box. In addition, the dialog box will respectively contain an information icon or an error icon.

Getting user response

You can use the DialogAsk function to obtain a user response. This function displays a dialog box containing a given message and two (or three) buttons with button text as given. The button3 argument is optional, and has to be tagged with the formal argument name. The return value is either 1, 2, or 3 and matches the button that was pressed.

Displaying progress messages

With the functions DialogProgress and StatusMessage you can provide progress information to the end-user of your model. The function DialogProgress will display a progress dialog box containing a message and (optionally) a progress meter under your control. The dialog box will disappear when you either call it with an empty message string, or when the execution from which it was called has ended. With the function StatusMessage you can display a message in the status bar.

Data entry dialog boxes

The following functions are available in AIMMS for scalar data entry dialog boxes.

- DialogGetString(*message,reference*[,*title*])
- DialogGetElement(*title,reference*)
- DialogGetElementByText(*title,reference,element-text*)
- DialogGetElementByData(*title,reference,element-data*)
- DialogGetNumber(*message,reference*[,*decimals*][,*title*])

- DialogGetPassword(*message,reference*[,*title*])
- DialogGetDate(*title,date-format,date*[,*nr-rows*][,*nr-columns*])

The *message* and *title* arguments must be strings, and *reference* must be a (scalar) reference to an identifier of the appropriate type. The *element-text* argument must be a string parameter defined over the set in which *reference* is an element parameter, whereas *element-data* must be a string parameter defined over the *reference* set plus a single additional simple set. The *decimals* argument must be a nonnegative integer.

AIMMS offers three basic functions for obtaining a string-valued, element-valued, or numerical, scalar value from the user, DialogGetString, DialogGet-Element and DialogGetNumber. These functions let you display a dialog box with your own message and dialog box title and an entry field for the scalar. AIMMS will only allow the user to enter a value that lies within the declared range of the scalar argument. The functions return 0 if the user presses the **Cancel** button, and 1 if the user presses the **OK** button.

Getting a user-supplied value

When displaying a data entry dialog box, the entry field displays the value of the scalar reference at the time of the call. You can use this to provide a default for the value that you want the user to supply. The default value in the function DialogGetNumber will be displayed with the number of decimal places as specified in the *decimals* argument.

Default value

When the elements in a set are not very descriptive to an end-user, AIMMS offers you an alternative way to have a user select a set element. With the function DialogGetElementByText you can supply an additional string parameter defined over the set from which you want the user to select an element. Instead of the element names, the dialog box will now display these descriptive texts.

Get element by description

Similarly, the function DialogGetElementByData displays string data divided into several columns from which the user can select a row corresponding to the desired element of a particular set. The function takes a two-dimensional string argument, defined over the set from which you want the user to select an element, plus an additional set of elements which will be used as column headers.

Description in multiple columns

With the function DialogGetPassword you can let the user enter a password. The function behaves like the function DialogGetString with the exception that the user-supplied input is not visible but shown as a sequence of '*' characters.

Obtaining password information

You can use the function DialogGetDate to let the end-user select a date that plays a role in your model, and store the resulting date in a string parameter. The *date-format* argument must be a date format specification string using the date- and time-specific components explained in Section 29.7.1 of the Language Reference. The *date* argument is an inout argument. If it contains a valid

Obtaining a date

date according to the specified format on input, AIMMS will set the initial date in the date selection dialog equal to the specified date. On output, the *date* argument contains the date selected by the user, according to the specified format.

Displayed number of months

With the optional *nr-rows* and *nr-columns* arguments of the DialogGetDate function, you can specify the number of rows and columns of months displayed in the date selection dialog (maximum 3 and 4 respectively, each with a default of 1). Thus, by specifying the maximum number of rows and columns you will be able to simultaneously display the days of 12 consecutive months within the dialog.

18.4.5 Data management functions

Controlled data management

The management of cases and datasets is a very important aspect of a successful decision support system. While the complete data management functionality is available to the end-user within the **Data Manager** or from the **Data** menu, you may want to have additional control over the data management process to perform special tasks.

Case functions

The following functions are available in AIMMS for performing case management tasks.

- CaseNew
- CaseFind(*case-path,case*)
- CaseCreate(*case-path,case*)
- CaseDelete(*case*)
- CaseLoadCurrent(*case*[,*dialog*])
- CaseMerge(*case*[,*dialog*])
- CaseLoadIntoCurrent(*case*[,*dialog*])
- CaseSelect(*case*[,*title*])
- CaseSelectNew(*case*[,*title*])
- CaseSetCurrent(*case*)
- CaseSave([*confirm*])
- CaseSaveAll([*confirm*])
- CaseSaveAs(*case*)
- CaseSelectMultiple([*cases-only*])
- CaseGetChangedStatus
- CaseSetChangedStatus(*status*[,*include-datasets*])
- CaseGetType(*case,case-type*)
- CaseGetDatasetReference(*case,data-category,dataset*)
- CaseCompareIdentifier(*case1,case2,identifier,suffix,mode*)
- CaseCreateDifferenceFile(*case,filename,diff-types*
      ,*absolute-tolerance,relative-tolerance,output-precision*)

The arguments *case, case1* and *case2* are element parameters in the predefined set AllCases, whereas *title, case-path* and *filename* are string arguments. The *status* argument, and the optional arguments *dialog, include-datasets* and *cases-only* must be 0 or 1. The optional *confirm* argument can be 0, 1 or 2. The default of both the *confirm* and *dialog* arguments is 1, *cases-only* is 0 by default. The arguments *case-type, data-category, dataset, suffix* and *mode* are elements of the sets AllCaseTypes and AllDataCategories, AllDatasets, AllSuffices and AllCaseComparisonModes, respectively, while the *diff-types* argument is an element parameter indexed over AllIdentifiers in the set AllDifferencingModes. The *absolute-tolerance, relative-tolerance* and *output-precision* arguments are numerical, scalar values.

The following functions are available in AIMMS for performing dataset management tasks.

Dataset functions

- DatasetNew(*data-category*)
- DatasetFind(*data-category,dataset-path,dataset*)
- DatasetCreate(*data-category,dataset-path,dataset*)
- DatasetDelete(*data-category,dataset*)
- DatasetLoadCurrent(*data-category,dataset*[,*dialog*])
- DatasetMerge(*data-category,dataset*[,*dialog*])
- DatasetLoadIntoCurrent(*data-category,dataset*[,*dialog*])
- DatasetSelect(*data-category,dataset*[,*title*])
- DatasetSelectNew(*data-category,dataset*[,*title*])
- DatasetSetCurrent(*data-category,dataset*)
- DatasetSave(*data-category*[,*confirm*])
- DatasetSaveAll([*confirm*])
- DatasetSaveAs(*data-category,dataset*)
- DatasetGetChangedStatus(*data-category*)
- DatasetSetChangedStatus(*data-category,status*)
- DatasetGetCategory(*dataset,data-category*)

The argument *data-category* must be an element parameter in the predefined set AllDataCategories, the argument *dataset* must be an element parameter in the set AllDatasets, whereas *title* and *dataset- path* are string arguments. The *status* argument, and the optional *dialog* argument, must be 0 or 1. The optional *confirm* argument can be 0, 1 or 2. The default of both the *confirm* and *dialog* arguments is 1.

The CaseNew, CaseLoadCurrent, CaseMerge, CaseLoadIntoCurrent functions, and their counterparts for datasets, have the same functionality as the corresponding items on the **Data** menu. With the optional *dialog* argument you can indicate whether you want a dialog box to be presented to the user. Without user input AIMMS will load the case or dataset provided as an argument of the function or return an error if no valid case or data set was provided. On its return, the case or dataset argument will contain the element of AllCases

Data menu functions

or AllDatasets associated with the case or dataset selected by the user. The return value can be:

0 : The user pressed the **Cancel** button.

-1 : An error occurred during importation.

1 : The data was loaded successfully.

Saving cases and datasets

The functions CaseSave, CaseSaveAll, CaseSaveAs, and their counterparts for datasets, offer the same functionality as the corresponding items on the **Data** menu. With the optional *confirm* argument you can specify whether you want the save to be confirmed by the user. The possible values are:

0 : never confirm,

1 : only confirm when required by the case, or

2 : always confirm.

The *confirm* argument defaults to 1. The return values of the save functions are as above.

Setting the default case type

When you do not want your end-users to select a case type when saving a new case in the case-save-as dialog box, you can preset the case type from within the modeling language through the predefined element parameter Current-DefaultCaseType. When this element parameter has a nonempty value, AIMMS will remove the case type drop-down list, and use the case type specified through CurrentDefaultCaseType.

Finding cases and datasets

With the functions CaseFind, CaseCreate, DatasetFind and DatasetCreate you can obtain the element of either the set AllCases or the set AllDatasets which is associated with a path to the indicated case or dataset. The functions CaseFind and DatasetFind will return 0 if no such case or dataset exists, while the functions CaseCreate and DatasetCreate will create nonexistent cases and datasets in exactly the same manner as if you were inserting new case or dataset nodes in the **Data Manager**.

Deleting cases and datasets

With the functions CaseDelete and DatasetDelete you can delete cases from the case and data category tree without using the AIMMS **Data Manager**. If you are deleting the active case or an active dataset, AIMMS will retain the data associated with that case or dataset, but remove its reference from the active case and dataset settings.

Loading cases and datasets

The functions CaseLoadCurrent and DatasetLoadCurrent load a case or dataset as active, and set the current case or dataset to the loaded data file. You can *import* a case or dataset into your current case through either the CaseMerge or CaseLoadIntoCurrent functions and their counterparts for datasets.

The functions CaseSetCurrent and DatasetSetCurrent let you set the current case or dataset *without loading any data*. As subsequent saves will save data in the current case or dataset, you should use these functions with care, and make sure that no data is inadvertently lost.

Setting the current case

With the functions CaseSelect, CaseSelectNew, DatasetSelect and DatasetSelectNew you can let the user select an existing or new case or dataset without actually opening it, or saving the current data to it. You can then further use this case, for instance, to import or export data using READ and WRITE statements.

Selecting a case or dataset

The function CaseSelectMultiple displays the **Multiple Cases** dialog box from the **Data** menu. With it, the user can select multiple cases from the case management tree, and use the selection for multiple case objects, calculations involving multiple cases, or creating your own batch run of cases. The selection made by the user is available to you through the predefined set CurrentCaseSelection. You can use it, for instance, to import selected data from all cases, and perform advanced case comparisons.

Selecting multiple cases

Sometimes you may want to check if the user has made changes to the data in the currently loaded case, or you may even want to change that status. The functions CaseGetChangedStatus and CaseSetChangedStatus do this. The status can be either 1 (case changed), or 0 (case unchanged). With the optional argument *include-datasets* you can indicate whether you also want to modify the status of all datasets included in the case. Similar functions are available for datasets.

Checking the case status

The function CaseCompareIdentifier can be used to compare two cases for a given identifier. For numerical identifiers this function returns the minimum, maximum, sum, average or total number of all data differences (depending on the *mode* argument). For non-numerical identifiers the total number of data differences is returned. To compare the data for all identifiers in a case at once and to dump the results in a text file you can use the function CaseCreateDifferenceFile. The resulting text file can then be used in a READ statement to apply the same differences to some other data instance.

Comparing identifier data

All data categories, datasets, case types and cases in an application are accessible in the model through a number of predefined sets and parameters. They are:

Case and dataset related identifiers

- the set AllDataCategories, containing the names of all data categories defined in the data manager setup window,
- the set AllCaseTypes, containing the names of all case types defined in the data manager setup window,

- the integer set AllDataFiles, representing all datasets and cases available with a particular project,
- the set AllDatasets, a subset of AllDataFiles, representing the collection of all datasets available in the project,
- the set AllCases, a subset of AllDataFiles, representing the set of all cases available for the project,
- the indexed element parameter CurrentDataset in AllDatasets and defined over AllDataCategories, containing the currently active datasets,
- the scalar element parameter CurrentCase in AllCases, and
- the scalar element parameter CurrentDefaultCaseType in AllCaseTypes.

Obtaining case type or data category

You can obtain the case type for each case through the function CaseGetType. For every dataset you can ask AIMMS to return its data category through the function DatasetGetCategory. With the function CaseGetDatasetReference you can, for every data category, obtain a reference to the dataset of that category included in the case. If no dataset is included, the *dataset* is set to the empty element, and the function returns 1. If an included dataset is nonexistent, the *dataset* is also set to the empty element, but the function now returns 0.

Data categories and case types

In the AIMMS **Data Manager**, data categories and case types are specified as a subcollection of identifiers from the model tree. Through the following functions you can obtain the contents of data categories and case types, should you need this information.

- DataCategoryContents(*data-category,identifier-set*)
- CaseTypeContents(*case-type,identifier-set*)
- CaseTypeCategories(*case-type,category-set*)

The argument *data-category* is an element of the set AllDataCategories. The argument *case-type* is an element of the set AllCaseTypes. The output arguments *identifier-set* and *category-set* must be subsets of AllIdentifiers and AllDataCategories, respectively.

Case type contents

The function CaseTypeContents will return a subset of identifiers which includes both the list of identifiers added to the case type itself, and the identifiers which are part of the data categories included in the case type. With the function CaseTypeCategories you can obtain the subset of data categories included in a case type, while the function DataCategoryContents returns the set of identifiers contained in a data category.

Obtaining datafile info

The mapping of the integer set AllDataFiles and its subsets onto the datasets and cases in the project is maintained by the **Data Manager**, and is not editable from within the model. Moreover, the numbering of cases and datasets may be different in every new session. During a session, however, the following functions give you access to all the information stored inside a datafile.

- DataFileGetName(*datafile*,*name*)
- DataFileGetAcronym(*datafile*,*acronym*)
- DataFileGetPath(*datafile*,*path*)
- DataFileGetDescription(*datafile*,*description*)
- DataFileGetTime(*datafile*,*time*)
- DataFileGetOwner(*datafile*,*user*)
- DataFileGetGroup(*datafile*,*group*)
- DataFileReadPermitted(*datafile*)
- DataFileSetAcronym(*datafile*,*acronym*)
- DataFileWritePermitted(*datafile*)
- DataFileExists(*datafile*)

The argument *datafile* is an element of the set AllDataFiles. The arguments *name*, *acronym*, *path*, *time*, *description*, *user* and *group* are string parameters. The time that a case or dataset was last saved will be returned as for ordinary files.

You can use the functions DataFileReadPermitted and DataFileWritePermitted to check whether a read or write action is permitted by the current user before actually performing that action. With the functions DataFileGetOwner and DataFileGetGroup you can obtain the user name and associated user group of the owner of the data file as they are stored in the datafile by AIMMS. More details about case and dataset security are contained in Section 20.4.

Checking security

Because the AIMMS data tree can be accessed by multiple users, some elements in the set AllDataFiles may refer to data files that have been removed by other users. Using the function DataFileExists you can check for the existence of a particular datafile referenced by an element of AllDataFiles. Thus, you can prevent your users from receiving error message about nonexistent data files, which may have little meaning to them.

Checking data file existence

To copy an existing case or dataset the following function is available.

Copying data files

- DataFileCopy(*datafile-src*,*acronym*,*datafile-dest*)

The arguments *datafile-src* and *datafile-dest* are elements of the predefined set AllDataFiles and *acronym* argument is a string parameter.

The import and export facilities in the AIMMS **Data Manager** allow you transfer parts of the case and dataset tree to another user, or vice versa. Whenever you want such imports or exports to take place automatically, for example when particular datasets must be imported on a regular basis, the following functions are available to perform such tasks from within your model.

Importing and exporting data files

- DataManagerExport(*filename,datafiles*)
- DataManagerImport(*filename*[,*overwrite*])
- DataImport220(*filename*)

The argument *filename* is a string parameter. The argument *datafiles* is a subset of the predefined set AllDataFiles. The optional argument *overwrite* can be either 0, 1 or 2, and defaults to 0.

Exporting data files

The function DataManagerExport exports the given set of data files to a newly created data manager file, deleting any previous contents. If the set *datafiles* contains cases with references to datasets which are not contained in *datafiles*, such datasets will also be exported. This ensures that any exported case refers to exactly the same data when imported by another user.

Importing data files

The function DataManagerImport imports *all* cases and datasets within the given data manager file into the current case tree. With the optional *overwrite* argument, you can specify AIMMS' behavior when any case or dataset in the import file already exists in the case tree. The following values are allowed:

0 : the end-user decides (default),
1 : existing entries are overwritten, or
2 : AIMMS creates new nodes if existing entries are present.

When AIMMS creates a new node alongside an existing entry, the name of the existing node is prefixed with the string 'Imported', followed by a number if there are multiple imported copies for the imported node.

Importing AIMMS 2.20 cases

The function DataImport220 allows you to import case files belonging to AIMMS 2.20 projects, which are incompatible with the new AIMMS 3 data storage scheme. You can use this function in an upgraded AIMMS 2.20 model, to upgrade cases created by end-users with your old AIMMS 2.20 project to the data manager tree of the (upgraded) AIMMS 3 project. The use of this function is especially useful when upgrading a case in your model requires additional data manipulation for example to store the label text of all set elements (which is no longer supported by AIMMS 3) in the original case file as string parameters in your AIMMS 3 model. The function returns 1 if the import succeeded, 0 if the user canceled the action, and −1 if the import failed.

18.4.6 Execution control functions

Execution control

During the execution of your AIMMS application you may need to execute other programs, delay the execution of your model, get the command line arguments of the call to AIMMS, or even close your AIMMS application.

The following execution control functions are available in AIMMS.

Control functions

- Execute(*executable*[,*commandline*][,*workdir*][,*wait*][,*minimized*])
- ShowHelpTopic(*topic*[,*filename*])
- OpenDocument(*document*)
- Delay(*delaytime*)
- ScheduleAt(*starttime*,*procedure*)
- ProjectDeveloperMode
- SessionArgument(*argno*, *argument*)
- ExitAimms([*interactive*])

The arguments *executable, commandline, workdir, filename, document, topic, starttime, argument* and *marker* are string arguments, the argument *delaytime* is a real number, while the arguments *wait, minimized* and *interactive* must be all either 0 or 1. The argument *procedure* must be an element of the predefined set AllProcedures. The argument *argno* must be an integer greater than or equal to 1.

With the Execute function you can start another application. You can optionally supply a command line argument for the application, indicate whether AIMMS should wait for the termination of the application, and whether the application should be started in a minimized state or not. As a general rule, you should not wait for interactive window-based applications. Waiting for the termination of a program is necessary when the program carries out some external data processing which is required for the further execution of your model. If you do not specify a working directory, AIMMS assumes that the current directory is the working directory.

The Execute function

Be aware that certain commands (such as "dir" or "copy") and features such as output redirection to file (">" or ">>") are executed by the DOS command shell rather than being executables themselves. If you want to make use of such features of the command shell, the application you call should be command.com (or cmd.exe if you want to make use of features of the Windows NT command shell) followed by the /c option to specify the specific command you want to be executed by the command shell, as illustrated in the following example.

Executing DOS commands

```
Execute( "command.com", "/c dir > dir.out" );
```

The function ShowHelpTopic starts up the help program with the indicated help file, and displays the requested topic. The function supports all of the help file formats described in Section 11.2. If you do not provide a particular help file, AIMMS will assume the default help file associated with your project.

The function ShowHelpTopic

The function
OpenDocument

The function OpenDocument opens the indicated document using the default viewer associated with the document extension. You can use it, for instance, to display an HTML file using the default web browser installed on a particular machine. The *document* argument need be a local file name, it could be a *URL* pointing to a page on the World Wide Web as well.

The Delay
function

With the Delay function you can block the execution of your model for the indicated delay time. You can use this function, for instance, when you want to change the particular slice of an identifier to be displayed on a page in the end-user interface at regular intervals. The delay time is specified in seconds.

The ScheduleAt
function

With the ScheduleAt function you can tell AIMMS that you want a particular procedure within your application to be run at a particular start time. The start time must be provided in the default format "*YYYY-MM-DD hh:mm:ss*". The function ScheduleAt will return immediately, while the indicated procedure will be run at the first opportunity after the given time when no other (interactive) execution is taking place. This form of scheduled execution is useful, for instance, when you want to initiate data retrieval from an external source at regular intervals.

The function
ProjectDevelop-
erMode

The function ProjectDeveloperMode lets you verify, from within your model, whether a project is run in developer mode or in end-user mode. In either case, you might want to perform different actions, e.g. activate a different menu, or open a different set of pages. The function returns 1 if the project is run in developer mode, or 0 otherwise.

The function
SessionArgument

When you open an AIMMS project from the command line, AIMMS allows you to add an arbitrary number of additional arguments directly after the project name. You can use these arguments, for instance, to specify a changeable data source name from which you want to read data into your model. With the function SessionArgument you can obtain the (string) value of argument *argno* ($\geq$ 1). The function fails if the specified argument number has not been specified.

The ExitAimms
function

To allow you to quit your application from within your model, AIMMS offers the function ExitAimms. Using this function, you can close your application without user intervention. You can optionally indicate whether the application must be closed in an interactive manner (i.e. whether the user must be able to answer any additional dialog box that may appear), or that the default response is assumed.

18.4.7 Debugging information functions

To help you investigate the execution of your model AIMMS offers several functions to control the debugger and profiler from within your model. In addition, a number of functions are available that help you investigate memory issues during execution of your model.

Debugging information

The following execution information functions are available in AIMMS.

Execution information functions

- IdentifierMemory(*identifier*[,*include-permutations*])
- MemoryStatistics(*filename*[,*append-mode*][,*marker-text*][,*show-leaks-only*] [,*show-totals*][,*show-since-last-dump*][,*show-mem-peak*][,*show-small-block-usage*])
- IdentifierMemoryStatistics(*identifier-set*,*filename*[,*append-mode*] [,*marker-text*][,*show-leaks-only*][,*show-totals*][,*show-since-last-dump*] [,*show-mem-peak*][,*show-small-block-usage*][,*aggregate*])

The argument *filename* is a string argument, while the arguments *include-permutations*, *append-mode*, *marker-text*, *show-leaks- only*, *show-totals*, *show-since-last-dump*, *show-mem-peak*, *show-small-block-usage* and *aggregate* must be all either 0 or 1. The argument *identifier* must be an element of the predefined set AllIdentifiers and the argument *identifier-set* must be a subset of this same set.

The function IdentifierMemory can be used to retrieve information about the total amount of memory that is occupied by a specific identifier. Its *include-permutations* arguments can be used to indicate whether or not permutations of the identifier should be included in the reported number. Permutations are used by AIMMS to quickly perform all kinds of operations over permuted instances of an identifier.

Displaying memory information

Through the function MemoryStatistics you can let AIMMS print statistics collected by AIMMS' memory manager to the specified file. Memory statistics will only be collected if the global AIMMS option memory_statistics is on. Through the *append-mode* option you can indicate whether output is to be appended to an existing file, or if any existing contents is to be overwritten. If multiple statistics are printed to the same file, you can specify a *marker-text* that will be printed at the top of statistics. All other optional arguments are additional settings to select specific types of statistics to be printed.

Displaying memory statistics

To restrict the collection of statistics to a subset of identifiers you can use the IdentifierMemoryStatistics. This function has two additional arguments, one to indicate the subset of identifiers subject to the collection process and one optional argument to indicate whether a individual report for every identifier in the set or a single aggregated report should be created.

Displaying identifier statistics

Profiler control   The following profiler control functions are available in AIMMS.

- ProfilerStart()
- ProfilerPause()
- ProfilerContinue()
- ProfilerRestart()

The profiling functions to control the AIMMS profiler allow for some extra flexibility over their use through the menu commands in the sense that it is easier to restrict profiling to a certain part of you model using these functions. The functions ProfilerStart, ProfilerPause and ProfilerContinue are available to start, pause and continue a profiler session. With the function ProfilerRestart the measumerment data of all statements and definitions is reset.

18.4.8   Obtaining license information

License
information
functions

The licensing functions discussed in this section allow you to retrieve licensing information during the execution of your model. Based on this information you may want to issue warnings to your end-user regarding various expiration dates, or adapt the execution of your model according to the capabilities of the license.

License
functions

The following licensing functions are available in AIMMS.

- LicenseNumber(*license*)
- LicenseStartDate(*date*)
- LicenseExpirationDate(*date*)
- LicenseMaintenanceExpirationDate(*date*)
- LicenseType(*type,size*)
- VARLicenseCreate(*var-license,license-number,module-code*[,*userdata*]
        [,*expiration-date*][,*days-left-warning*][,*number-of-users*][,*is-network*])
- VARLicenseExpirationDate(*var-license,date*)
- AimmsRevisionString(*revision*)

All arguments are strings. The *var-license* argument is input, all other arguments are output arguments, except for the function VARLicenseCreate which has only input arguments.

License number   Through the function LicenseNumber you can retrieve the license number of the currently active AIMMS license. It will return a string such as "015.090.010.007" if you are using an AIMMS 3 license, or a string such as "1234.56" if you are using an AIMMS 2 license.

You can use the functions LicenseStartDate, LicenseExpirationDate and LicenseMaintenanceExpirationDate to obtain the start date, expiration date and maintenance expiration date of the currently active AIMMS license, respectively. All dates will be returned in the format "YYYY-MM-DD". If a particular date has not been specified in the AIMMS license, AIMMS will return "No start date", "No expiration date" or "No maintenance expiration date", respectively.

License dates

The function LicenseType will return type and size information of the currently active AIMMS license. Upon success, the *type* argument contains the license type description (e.g. "Economy") and the *size* argument contains a description of the license size (e.g. "Large").

License type

Through the function VARLicenseCreate you can create a VAR license to be used with your project. In the *var-license* argument you must specify the name of the VAR license (file), as you have specified it in the **Project Security** dialog box, or in the attribute window of the main model node or a section of your model. The *license-number* argument must be specified in the format "015.090.010.007", and the *expiration-date* argument in the format "YYYY-MM-DD". All other arguments directly correspond to the input fields in the **VAR License Manager** dialog box discussed in Section 20.2.

VAR license creation

Through the function VARLicenseExpirationDate you can obtain the expiration date of a VAR license that is used within your project. In the *var-license* argument you must specify the name of the VAR license (file), as you have specified it in the **Project Security** dialog box, or in the attribute window of the main model node or a section of your model. The expiration date will be returned in the format "YYYY-MM-DD". You can find more information about VAR licensing in Section 20.2.

VAR license expiration

You can use the function AimmsRevisionString if you want to obtain the revision number of the currently running AIMMS executable. The revision string returned by the function has the format "x.y.b" where x represents the major AIMMS version number (e.g. 3), y represents the minor AIMMS version number (e.g. 0), and where b represents the build number (e.g. 476) of the current executable. You can use this function, for instance, to make sure that your end-users use an AIMMS version that is capable of certain functionality which was not available in earlier AIMMS releases.

AIMMS *revision*

Chapter 19

Calling AIMMS

This chapter

This chapter discusses the command line options of the AIMMS program, and explains the details for running AIMMS end-user applications. In addition, the chapter explains how you can link AIMMS to your own program as a DLL, and presents a short overview of the functionality available through the AIMMS-specific Application Programming Interface (API) provided by this DLL.

19.1 AIMMS command line options

Calling AIMMS

On the AIMMS command line, you can specify a number of options and arguments that will influence the manner in which AIMMS is started. The following line illustrates the general structure of a call to the AIMMS program.

aimms.exe *[command-line-options] [project-file [session-arguments]]*

Command line options

Table 19.1 provides an overview of the command line options that you can specify. AIMMS offers both long and short option names, and some options require a single argument. All short option names start with a single minus (-) sign, followed by a single character. By convention, short options that require an argument use capital characters. The long option names are always preceded by a double minus sign (--), followed by a descriptive text. In general, the long option names are easier to remember, while the short names permit a more compact command line. Short option names without an argument may be appended one after another with only a single minus sign at the beginning.

Specifying a user

When an AIMMS project is linked to an end-user database (see Chapter 20), you must log on to the project before being able to run it. Through the *--user* command line option, you can specify a user name and optionally a password with which you want to log on to the system. When you specify just a user name, a log on screen will appear with the provided user name already filled in. If you specify a password as well, AIMMS will verify its correctness and skip the log on screen altogether if the user name- password combination is acceptable. Providing both the user name and the password is not recommended for interactive use, but may be convenient when you want the model to run unattended.

Long name	Short name	Argument
--user	*-U*	user[:password]
--data	*-D*	data manager file
--backup-dir	*-B*	backup directory
--log-dir	*-L*	log directory
--config-dir	*-C*	configuration directory
--license		license name
--run-only	*-R*	procedure name
--minimized	*-m*	—
--maximized	*-x*	—
--hidden		—
--as-server		—
--developer	*-d*	—
--end-user	*-e*	—
--user-database		—
--no-solve		—
--help	*-h*	—

Table 19.1: AIMMS command line options

By default, AIMMS gets its cases and datasets from the last selected project-dependent data manager file. You can always select your own choice of data manager file using the **File-Open** menu. With the *--data* flag, you can specify an alternative data manager file with which you want to open the project. You can only use this option if you also specify a project file.

Specifying a data manager file

With the *--backup-dir* and *--log-dir* options you can override the default directories where AIMMS will store temporary information such as case and model backups, the AIMMS and solver listings, and the message log. You can modify the defaults for these directories using the project options dialog box (see Section 21.1).

Backup and log directories

By default, AIMMS stores a number of global configuration files, such as the AIMMS license file, VAR license files and the solver configuration file, in the common application area of your computer (see also Section 2.6.4). If you want to store configuration files in a different location, you can indicate this through the *-- config-dir* option. You can use this option, for instance, to indicate where the configuration files for your particular machine can be found when the AIMMS system that you use is stored on a network disk, and when you do not use a license server.

AIMMS configuration

Through the *--license* option you can select any AIMMS license that you installed in the AIMMS **License Configuration** dialog box (see also Section 2.6).

License name

The value that you specify for the -- *license* option should match the name of a license in the left pane of the **License Configuration** dialog box.

User database
location

When your application has been set up for use by multiple users, all user and group information associated with the application is stored in a separate (encrypted) user database (see Section 20.3 for more details on this topic). Through the *--user-database* option you can move the location of this user database file (to for example a single location that is shared among all users on the network) even though you might not have developer rights to the application.

Running
minimized,
maximized,
hidden, or as
server

Through the *--minimized, --hidden* and *--maximized* options you can indicate whether you want AIMMS to start in a minimized or hidden state (i.e. just as a button on the task bar, or not visible at all), or to fill up the entire screen. Running AIMMS minimized or hidden may be convenient when AIMMS is called non-interactively from within another program through the AIMMS API (see Chapter 30 of the Language Reference). In this way, your program can use AIMMS to solve an optimization model after which it resumes its own execution. The --as-server option extends the --hidden option, and should be used when AIMMS is started with limited privileges by a system service (e.g. through the Internet Information Server). It suppresses all dialog boxes that may appear during startup of AIMMS, as well as during the execution of your model.

Developer
versus end-user
mode

With the *--developer* and *--end-user* options you can request AIMMS to start up a project in developer mode or end-user mode, respectively, overriding the default start-up mode of the project (see also Section 19.2). When you are the developer of an AIMMS- based application, you are always allowed to run the application in end-user mode. If you are using a VAR-licensed AIMMS application, starting the application in developer mode may be prohibited, or subject to an additional developer password (as explained in Section 19.2).

Solverless
AIMMS *sessions*

AIMMS strictly enforces that the number of AIMMS sessions with full solving capabilities running on your computer simultaneously is in accordance with your AIMMS license. Typically, for a single-user license, this means that you can you can only start up a single AIMMS session that is capable of solving optimization programs at a time. However, for every fully capable AIMMS session, AIMMS also allows you to start up an additional AIMMS session without solving capabilities. You can use such a session, for instance, to make modifications to your model, while a first session is executing an optimization run. In that case, AIMMS will present a dialog box during start up to indicate that the session has no solving capabilities. You can suppress this dialog box, by specifying the *--no-solve* command line option.

When you want to run an AIMMS project unattended, you can call AIMMS with the *--run-only* option. This option requires the name of a procedure in the model, which will be executed after the project is opened. When you use the *--run-only* option, all other initial project settings, such as the initial case, procedure and page settings (see Section 19.2), will be ignored. AIMMS will, however, call the procedures MainInitialization and MainTermination as usual. Once the procedure has finished, the AIMMS session will be terminated. You can only specify the *--run-only* option if you also specify a project file on the command line.

Executing a procedure and terminating AIMMS

AIMMS will interpret the first non-option argument on the command line as the name of the project file with which you want to open AIMMS. If you specify a project file, the settings of the project may initiate model-related execution or automatically open a page within the project.

Opening a project to run

If you want to open a project for editing purposes only, you should hold down the **Shift** key when opening the project. The initial actions will also not be performed if the command line contains the *--run-only* option. In this case execution takes place from within the specified procedure only.

Opening a project to edit

Directly after the name of the project file, AIMMS allows you to specify an arbitrary number of string arguments which are not interpreted by AIMMS, but can be used to pass command line information to the project. In the model, you can obtain the values of these string arguments one at a time through the predefined function SessionArgument, which is explained in more detail in Section 18.4.6.

Passing session arguments

The following call to AIMMS, will cause AIMMS to start the project called transport.prj in a minimized state using the user name batchuser with password batchpw, run the procedure ComputeTransport, and subsequently end the session. A single argument "Transport Data" is provided as a session argument for the model itself.

Example

```
aimms --minimized --user batchuser:batchpw --run-only ComputeTransport \
    transport.prj "Transport Data"
```

Note that the \ character at the end of the first line serves as the continuation character to form a single command line. Using the short option names, you can specify the same command line more compactly as

```
aimms -mUbatchuser:batchpw -RComputeTransport transport.prj "Transport Data"
```

In this command line, the -m and -U options are combined. No space is required between a short option name and its argument.

Using session
arguments

Given the above AIMMS call, you can use the function SessionArgument to fetch the first session argument and assign it to the string parameter ODBCDataSource as follows.

```
if ( SessionArgument(1, ODBCDataSource) ) then
    /*
     *  Execute a number of READ statements from ODBCDataSource
     */
endif;
```

Following this statement, the string parameter ODBCDataSource will hold the string "Transport Data". In this example, the string parameter ODBCDataSource is intended to serve as the data source name in one or more DATABASE TABLE identifiers, from which the input data of the model must be read.

19.2 Running end-user applications

Running
end-user
projects

An AIMMS project can run in two different modes, *developer* mode and *end-user* mode. While the developer mode allows you to use the full functionality described in this User's Guide, the end-user mode only allows you to *access* the end-user pages of the AIMMS project that were created in developer mode.

Disabled
functionality

The AIMMS end-user mode lacks the essential tools for creating and modifying model-based applications. More specifically, the following tools are not available in end-user mode:

- the **Model Explorer**,
- the **Identifier Selector**,
- the **Page Manager**,
- the **Template Manager**,
- the **Menu Builder**, and
- the **Data Management Setup** tool.

Thus, in end-user mode, there is no way in which an end-user can modify the contents of your AIMMS-based application.

Allowed usage

AIMMS end-users can only perform tasks specified by you as an application developer. Such tasks must be performed through data objects, buttons and the standard, or custom, end-user menus associated with the end-user pages in your project. They include:

- modifying the input data for your model in the end-user interface,
- executing procedures within your model to read data from an external data source, or performing a computation or optimization step,
- viewing model results in the end-user interface,
- writing model results to external data sources or in the form of printed reports, and

■ performing case management tasks within the given framework of data categories and case types.

Thus, an end-user of your application does not need to acquire any AIMMS-specific knowledge. The only requirement is that the interface that you have created around your application is sufficiently intuitive and clear.

Before you can distribute your AIMMS model as an end-user application, two requirements have to be fulfilled:

Requirements

■ you must ensure that your modeling application starts up in end-user mode, either using the **Options** dialog box (see Section 21.1) or by *VAR licensing* your application (see Section 20.2), and
■ you need to associate a *startup page* with your application which will be displayed when your application is started by an end-user.

For every end-user project, you must associate a single page within the project so that it becomes the project's *startup page*. Such an association can either be made directly by selecting a page for the 'Startup Page' option in the AIMMS **Options** dialog box (see Section 21.1), or implicitly as the first opened page in the *startup procedure* of the project using a call to the PageOpen function.

Assigning a startup page

After opening your project in end-user mode, AIMMS will display the startup page. As all communication between the end-user and your model is conducted through end-user pages of your design, this first page and/or its menus must provide access to all the other parts of your AIMMS application that are relevant for your end-users. If all pages are closed during a session, the end-user can still re-open the startup page using the first page button ⊞ on the **Project** toolbar, or via the **View-First Page** menu.

Role of startup page

In addition to a startup page you can also provide a startup procedure in the project-related AIMMS options. Inside the startup procedure you can perform any initializations necessary for an end-user to start working with the project. Such initializations can include setting up date or user related aspects of the project, or reading the data for particular identifiers from a database.

Startup procedure

By default, AIMMS will display a splash screen during startup. When you are opening AIMMS with a particular project, you can replace AIMMS' own splash screen with a bitmap of your choice. If the project directory contains a bitmap (.bmp) file with the same name as the project file, AIMMS will display this bitmap file on the splash screen. In such a bitmap you can display, for instance, version information about your project.

Replacing the splash screen

19.3   Calling AIMMS as a DLL

Use AIMMS *as a DLL*

In addition to starting the AIMMS program itself, you can also link AIMMS, as a DLL, to your own application. Using AIMMS as a DLL has the advantage that, from within your program, you can easily access data with AIMMS and run procedures in the associated AIMMS project. Thus, for instance, when your program requires optimization, and you do not want to bother writing the interface to a linear or nonlinear solver yourself, you can

- specify the optimization model algebraically in AIMMS,
- feed it with data from your application, and
- retrieve the solution after the model has been solved successfully.

Use the AIMMS *API*

AIMMS offers an extensive Application Programming Interface (API) through which you can

- view and modify the contents of simple and compound sets in an AIMMS model,
- view and modify the contents of scalar and multidimensional parameters and variables, and
- run procedures within a model (synchronously or asynchronously).

All data communication of multidimensional identifiers between AIMMS and the external program is performed in a sparse manner, i.e. only tuples with a nondefault value are passed. For further details about the AIMMS API you are referred to the AIMMS Language Reference.

Programming required

Through the AIMMS API you have complete control over the data inside your model. Use of the AIMMS API requires, however, that you set up the interface to your model in a programming language such as C/C++ or Fortran. While the control offered by the AIMMS API may be relevant for advanced or real-time applications where efficiency in data communication is of the utmost importance, the learning curve involved with mastering the API may be too long when you only want to perform simple tasks such as communicating data in a blockwise manner and running procedures inside the model. In such cases you might consider setting up the communication using either ASCII data files or databases.

Chapter 20

Project Security

When you are creating a model-based end-user application there are a number of security aspects that play an important role.

Project security

- How can you protect the proprietary knowledge used in your model?
- How can you prevent the end-users of your application from modifying the project (thereby creating a potential maintenance nightmare)?
- How can you distinguish between the various end-users and their level of authorization within your application?

AIMMS offers several security-related features that address the security issues listed above. These features allow you to

This chapter

- irreversibly encrypt the source code of your model,
- password-protect, (reversibly) encrypt, and/or license the project and model files that are part of your application,
- introduce authorization levels into your model, and
- set up an authentication environment for your application.

This chapter describes these mechanisms in full detail, together with the steps that are necessary to introduce them into your application.

20.1 One way encryption

AIMMS supports two manner of encryption of your model source. It is up to you to choose the encryption scheme that works for you.

Two ways of encryption

- If your only concern is to protect your investment in model development, but do not need on-site access to your model and do not want to license (parts) of your model strictly coupled to an AIMMS license number, the easiest way to accomplish this protection is to use the encryption scheme discussed in this section.
- If you do need on-site access to your model and/or want to protect your model by strictly coupling it to an AIMMS license number, you should use the more involved VAR licensing scheme discussed in Section 20.2.

*One way
encryption*

By using AIMMS' one way encryption scheme you simply produce a version of *all* model files in your model that are irreversibly encrypted. You can create a one way encrypted version of your model through the **File-Export** menu, which will open the **Project Export** dialog box illustrated in Figure 20.1.

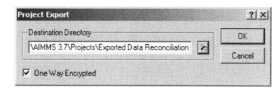

Figure 20.1: The **Project Export** dialog box

*Exporting your
project*

In the **Project Export** dialog box, you are offered the possibility to export all the files in your project to an export directory of your choice, ready to be shipped to your customer(s). By checking the **One Way Encrypted** check box, all .amb model files to which you have developer access, will be replaced by equivalent .aeb (AIMMS encrypted base) files.

*No access to one
way encrypted
models*

All the model source in the .aeb files is irreversibly encrypted, and access to the model tree is unconditionally prohibited in a one way encrypted project.

20.2 VAR licensing project components

*Protecting your
investment*

If you have invested considerably in the development of an AIMMS model for a particular application area, it is not unreasonable that you should want to protect your investment and strictly control the usage of your application. To support you in this task, AIMMS allows you to password-protect the access to, encrypt, and license the use of individual project and model files through AIMMS' VAR licensing scheme. Through VAR licenses you can strictly enforce your licensing requirements, by coupling the use of your project to individual AIMMS license numbers issued to your customers.

Protect a project

Through the **Settings-Project Security** menu you can set up the protection of your project and model files. It will open the **Project Security** dialog box illustrated in Figure 20.2. If you want to password-protect or license your project or model as a whole, completing the appropriate sections of the **Project Security** dialog box will suffice. If you want to license separate sections of your model tree containing a fully functional source module ready for use by third parties, you must set up the licensing directly in the model tree, as explained in Section 20.2.2.

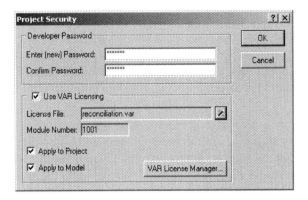

Figure 20.2: The **Project Security** dialog box

The simplest form of project security is by protecting your project and model file through a *developer password*. You can add a developer password to your project by completing the **Developer Password** section in the **Project Security** wizard. An existing developer password is removed by entering an empty password. Once a project is password protected, you need to enter this password every time you open the project in developer mode.

Password protection

Note that adding a developer password to a project will not encrypt the information stored in the model file per se. Although the model file stores its information in a binary format, parts of it may still be readable, potentially giving away proprietary information. To encrypt the information in the model file, you should protect it by a VAR license.

No encryption

AIMMS will add the developer password to both the project and the model file. To open a project in developer mode successfully, the developer password you enter must match the passwords stored in both files. If the password you enter does not match either (stored) password, full developer access to the project will be denied.

Project and model password must match

Through the **Use VAR Licensing** section of the **Project Security** dialog box, you can protect the use of an AIMMS-based modeling application by means of a VAR license. With such a license you can impose restrictions on

Licensing a project

- the expiration date of the application,
- whether the license is a stand-alone or a network license, and
- in the case of a network license, the number of network users that can run the application concurrently.

Before you can create VAR licenses yourself, you must register with Paragon Decision Technology (PDT) as a Value Added Reseller (VAR). After you have done so, your AIMMS license will be extended with a unique VAR identifica-

VAR identification code

tion code, which will enable the **VAR License Manager** in AIMMS (see Section 20.2.1). Using this tool you are able to create the VAR licenses necessary to protect your own AIMMS-based applications on the basis of your VAR identification code. The uniqueness of the VAR identification code issued by PDT ensures that licenses to protect your applications can only be created by you.

Module
identification
code

In addition to protecting a project with your own unique VAR identification code, you can associate a unique (integer) module identification code with a particular project (or source module) itself. AIMMS will only allow the use of a particular module if *both* the VAR and module identification codes stored in the project and/or model files coincide with the module identification code stored in the VAR license. In this manner, you can license several projects and source modules independently.

Existing VAR
license file

In the **Project Security** dialog box, you can license the use of your AIMMS-based application on the basis of an existing VAR license file. If you have not yet created a VAR license file, you can open the **VAR License Manager** through the equally named button on the **Project Security** dialog box. By selecting a VAR license file using the **License File** wizard ![icon], AIMMS will read the associated module identification code from the selected license file (if applicable) and display it in the **Module Number** field.

License project
and/or model

You can license the use of the project file and the model file independently. If you license the use of the model file associated with your project, AIMMS will just add the appropriate licensing attributes to the main model node in the model tree (as explained in Section 20.2.2). If required, you can later modify these attributes manually to suit your particular needs.

Developer
access

You can only obtain developer access to a VAR licensed project and/or model under the following conditions:

- your AIMMS license contains a unique VAR identification code, and
- this VAR identification code matches the VAR identification codes stored in the project or model file.

End-user access

If the above two conditions are not met, AIMMS will still grant end-user access if the project is accompanied by a VAR license file which

- has not expired, and
- matches the AIMMS license with which the project is run.

If AIMMS cannot find such a VAR license file, end-user access to the project (or module) is denied.

If you have prepared multiple VAR license files each associated with a single AIMMS license, you can distribute your project along with all generated VAR license files as a single package by using AIMMS' *VAR license directory* facility. Instead of specifying a VAR license *file* in the **License File** attribute, AIMMS also allows you to specify a *directory* for this attribute. In that case, AIMMS will look inside the specified directory for a VAR license file (with the .var extension) that matches the AIMMS license number. For instance, if the AIMMS license number is 15.90.10.7, AIMMS will look inside the specified directory for a VAR license file called 015090010007.var.

VAR license directory

20.2.1 Creating a VAR license

When your AIMMS license contains a unique VAR identification code, the **Tools-License-VAR License** menu will be enabled. Through this menu, AIMMS allows you to generate you own VAR licenses. It will open the **VAR License Manager** dialog box illustrated in Figure 20.3.

Creating a VAR license

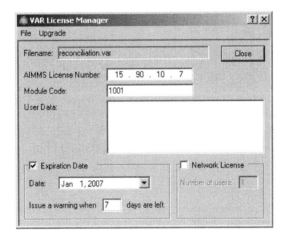

Figure 20.3: The **VAR License Manager**

In the **VAR License Manager**, you can enter the AIMMS license number of the end-user to whom you intend to distribute your module. This uniquely couples your VAR license to a single AIMMS license. To make your module available to a group of users, who all have similar AIMMS license numbers, you can replace the varying part of these AIMMS licenses by zero. When you enter a zero as one of the components of the AIMMS license number (all of which are separated by dots), AIMMS will accept any number for that component of the AIMMS license. Thus, if you license your module to license number 0.0.0.0, you permit your application to be used by *any* AIMMS license, while a VAR license with license number 15.90.10.0 allows your application to be used by all AIMMS licenses starting with 15.90.10.

AIMMS *license number*

Module code   The module code that you specify for a particular VAR license must coincide with the module identification code stored in the project and/or model files associated with the module that you want to license. AIMMS will only allow an end-user to use your module if the module codes in the VAR license and in the module-related files are the same.

Expiration date   In addition to coupling your module to a single AIMMS license (or group of AIMMS licenses), you can also limit the period of its use. When you enter an expiration date in the **Var License Manager**, AIMMS will not allow any use of the module after that date. By default, AIMMS will warn your end-user about the expiration date if the application is started within 7 days of the expiration date.

Stand-alone versus network   AIMMS and VAR licenses can be stored on a stand-alone machine, as files, or can be provided by an AIMMS network license server to your local area network. You can turn a VAR license into a network license by checking the **Network license** check box and providing the number of concurrent users. In this case, the AIMMS license number must be the license number of the corresponding AIMMS network license.

User data   If your licensing needs go beyond the standard licensing features described above, AIMMS allows you to store a user-definable string in a VAR license file in which you can store whatever information you require for your particular licensing scheme. When a license has been activated for a particular module, this user data can be made available through a locally declared string parameter (see Section 20.2.2). In the **User data** area of the VAR license manager you can enter any string, of maximum 256 characters, that you want to pass on to your module.

Opening and saving VAR licenses   With the **Open**, **Save** and **Save as** items in the **File** menu of the **VAR License Manager** you can open an existing VAR license for modification, resave it, or save a license in a new VAR license file. When your end-users include your module into their model, the VAR license file must be available, either as a local file or through the AIMMS network license server, under the name that you specified in the **License File** field of the **Project Security** dialog box.

20.2.2 Licensing model sections

Multiple model files   As explained in Section 4.2, AIMMS allows you to associate a separate source file with every subtree of your model. Such a separation is not only useful in a multi-developer environment, but also for the separate storage of those parts of your model that can be considered as more or less independent modules. For instance, such a module could consist of

■ the AIMMS interface to a library of DLL functions providing functionality that is not easily modeled in AIMMS itself, or

■ a model with well-defined input data to solve a particular problem class.

To create a licensed module, you must first store the source code of the section containing the module in a separate source file by completing the **Source file** attribute of the section. This step is not necessary if you want to VAR license an entire model. When you open the attribute form of the main model node or of the separated section, it will contain a number of additional attributes, as illustrated in Figure 20.4. These attributes allow you to turn the main model

Creating a licensed module

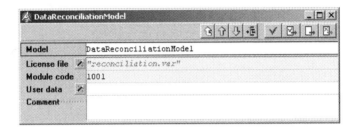

Figure 20.4: Licensing attributes of the main model node

node (or the section) into a licensed module. *Note that these additional licensing attributes are only visible when your* AIMMS *development license contains a unique VAR identification code.* Without the VAR identification code, licensing of a source module is not possible.

To license a source module, you must select the name of a license file that you want to associate with the module through the **License file** wizard ![icon] AIMMS will automatically enter the appropriate **Module code** if applicable. If you enter a string parameter in the **User data** attribute, AIMMS will assign the user data stored in the VAR license file to that string parameter for further use in the module.

Licensing attributes

After you have turned a section into a licensed source module, AIMMS will automatically encrypt the associated source file, making it impossible for your end-users to read its contents when you distribute it. In addition, when an end-user includes your licensed module in his model tree, the subtree containing your code can no longer be opened. These restrictions do not apply to you as the developer of the module.

Module encryption

20.3   User authentication

User authentication

When an application is set up for use by multiple users, it is usually considered desirable that users have access to only those parts of the application that are of interest to them, and can be given or denied the right of access to each others data. AIMMS allows you to set up such a controlled environment around your model-based application. This section describes the security features available in AIMMS.

20.3.1   Introduction

Users and groups

In a multi-user environment, a log on procedure is commonly employed to identify and authenticate the particular user who wants to make use of a system at a particular time. Users can own distinct resources within the system, and can control the access of other users to such resources. In addition, this scheme is often extended using the notion of *user groups* to categorize users who share a certain characteristic (e.g. who work in the same department), and for that reason should be able to access each others data.

Access rights

Complementary to the distinction of users and user groups and their associated rights to access *data*, is the question of which rights should be assigned to a specific user in accessing particular *functionality* within a system. For instance, in an AIMMS application, one might want to restrict the access to particular end-user pages, not allow a user to make changes to the values of certain identifiers within an end-user page, or disable his ability to execute particular parts of the model.

End-user roles

Rather than defining these access rights for every individual user, or for every user group, at a particular installation site, it often makes more sense to distinguish the several *roles* an end-user can play within an application, and link the access rights of a user to his role within the application. The number of roles that need to be defined for a particular application and their associated level of authorization, is usually fixed and relatively small.

Security in AIMMS

To help you set up a flexible environment for providing security to your model-based application, AIMMS supports the concepts of authorization levels, model users and user groups as discussed above. To help you accomplish this task, AIMMS provides a number of security tools for the definition of authorization levels during application development, as well as for adding users and user groups once the application is installed at a particular end-user site.

Although authorization levels, users and user groups all play a role in securing an application, the responsibilities for their creation, use and administration are quite different. *Responsibilities*

- The creation and change of authorization levels can only be carried out if the AIMMS project is opened in developer mode, as the set up of authorization levels with their associated rights is part of the design of a model-based application.
- The creation and modification of users and user groups is a task for a site-specific user administrator, and can also be performed if the project is opened in end-user mode.

20.3.2 Setting up and using authorization levels

You can associate authorization levels with your modeling application through the **Settings-Authorization Level Setup** menu, which is only available in development mode. It will open the **Authorization Level Manager** illustrated in Figure 20.5. In this dialog box you can add new authorization levels to your application by adding nodes to the list of existing authorization levels. *Setting up authorization levels*

Figure 20.5: The **Authorization Level Manager**

Password protection
By default, during log on end-users of a protected AIMMS application will obtain the authorization level that has been assigned to them by the (local) user administrator. For every authorization level in your application you can specify a password that allows end-users to obtain an authorization level different from their default level during an AIMMS session. By double- clicking on an authorization level (or through the **Edit-Properties** menu) you open the **Properties** dialog box displayed in Figure 20.6. In the **Password** tab of this dialog

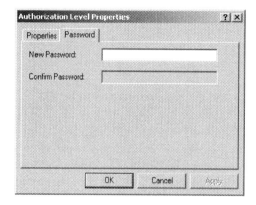

Figure 20.6: The authorization level **Properties** dialog box

box you can specify the password required to switch to this authorization level during a session (see also Section 20.3.4). If you do not specify a password, end- users can switch to that authorization level as long as they have access to the **File-Authorization** menu.

Default authorization level
Whenever you have authorization levels in your AIMMS project, one level is designated as the *default* level. In the authorization level tree, the default level will be shown in bold. Initially, AIMMS will make the first authorization level that you add to the tree the default level. You can modify the default authorization level using the **Edit-Set Default** menu. AIMMS uses the default authorization level for users to whom no authorization level has been associated (see also Section 20.3.3).

Use in the model
Within an AIMMS model, you have access to all the authorization levels defined for the associated project through the predefined set AllAuthorizationLevels. In addition, the currently active authorization level is available through the predefined element parameter CurrentAuthorizationLevel in the set AllAuthorizationLevels. The value of this element parameter changes, whenever a user logs on to the application, or changes the authorization level during a session.

Using the predefined set and element parameter discussed above, you can set up your own customized authorization level based security scheme within your application. By defining your own subsets of, and parameters over, the set AllAuthorizationLevels you can specify conditions to check whether the current user is allowed to perform certain actions.

Authorization-based security

Assume that ExecutionAllowed is a two-dimensional parameter defined over the set AllAuthorizationLevels and a user-defined set of ActionTypes. Then the following code illustrates the use of the element parameter CurrentAuthorizationLevel to allow or forbid a certain statement to be executed.

Example

```
if ( ExecutionAllowed(CurrentAuthorizationLevel, 'Solve') ) then
    solve OptimizationModel;
else
    DialogError( "Your authorization level does not allow you\n" +
                 "to solve the optimization model" );
endif;
```

You can also use parameters defined over AllAuthorizationLevels to influence the appearance and behavior of the end-user interface. More specifically, the following aspects of an AIMMS end-user interface can be influenced through the nonzero status of (indexed) parameters:

Use in the interface

- the access to a page through the page tree-based navigational controls,
- the visibility of graphical (data) objects on a page,
- the read-only status of data in a data object, and
- the visibility and enabled/disabled status of menu items and buttons.

If such parameters are defined over AllAuthorizationLevels, these aspects can be directly linked to the permission appropriate for a specific authorization level by slicing over the element parameter CurrentAuthorizationLevel.

20.3.3 Adding users and groups

All user and group information associated with a particular AIMMS application is stored in a separate (encrypted) user database file. Before you can start adding users and user groups you must first link your application to an existing user database or create a new one. As users and groups are site rather than model specific, all user management tasks can be performed from within both development and end-user mode of a project.

User databases

You can link to an existing user database, or create a new one, through the **Settings-User Setup-Link** menu. This will open a dialog box to let you select an existing or new user database file, with the .usr extension. If you select an existing user database which is password protected, AIMMS will only allow you to link to the user database after entering the correct password.

Linking to a user database

Unlinking a user database

Through the **Settings-User Setup-Unlink** menu you can unlink a linked user database. If the user database is password protected, you can only unlink after entering the correct password. Thus, you can effectively prevent your end-users from circumventing the authentication procedure by unlinking the user database.

Editing a user database

You can edit a user database linked to your application through the **Settings-User Setup-Edit** menu. After a password-check (if the user database is password protected), AIMMS will open the **User Manager** window illustrated in Figure 20.7.

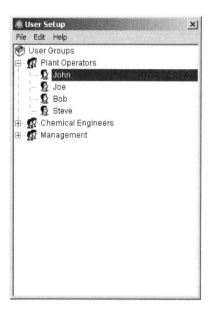

Figure 20.7: The **User Manager** window

Password protecting a user database

You can add a new user administrator password or change an existing one via the **File-Change Password** menu in the **User Manager**. Adding a user administrator password has the following effects that linking, unlinking and editing the user database is password-protected.

Adding users and groups

You can add new user groups and users as new nodes in the user manager tree. As each user must be a member of a unique user group, you must first add one or more user groups to the user manager tree before you can add users. Figure 20.7 illustrates a user group and user configuration. Insertion, deletion and modification of user and group nodes within the user manager tree is carried out in the usual fashion (see also Section 4.3).

The user group in which you position a user will become the *default user group* of that user. When a user logs on to an AIMMS application, he will automatically become a member of his default group. During a session, group membership can be modified through the **File-Authorization-Group** menu (see also Section 20.3.4). Group membership is only relevant in determining the access rights to case data (see also Section 20.4).

Default user group

The user manager in AIMMS lets you set up a hierarchical group structure. You can use it to set up a hierarchical protection scheme for case data by assigning the relevant access rights to members of parent and child groups (see Section 20.4).

Hierarchical group structure

For every new user or user group added to the user database, you can set its properties in the associated **Properties** dialog box illustrated in Figure 20.8. You can open the dialog box by double clicking on a user or user group, or

User and group properties

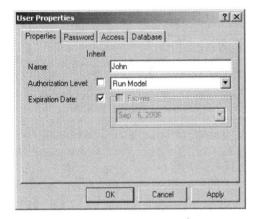

Figure 20.8: The user or user group **Properties** dialog box

through the **Edit-Properties** menu. In the dialog box you can specify properties such as

- the authorization level associated with an account, its expiration date and password,
- whether the user or user group concerned is allowed to enter the project in development mode,
- the default access rights for cases and datasets (see Section 20.4), and
- the default ODBC or OLE DB name and password associated with a user or group of users.

Expiration date   In the **Expiration Date** field in the **Properties** dialog box of a user account you can select an expiration date for the account using a calendar control. To remove an existing expiration date, you should uncheck the **Expires** check box. AIMMS will not allow the user to log on to the application when his account has expired.

Authorization level   In the **Authorization Level** field in the **Properties** dialog box of a user or user group you can enter the authorization level assigned to that user or user group. When you assign an authorization level to a user group, newly created user groups and users within that group will automatically inherit its authorization level. When you assign a default authorization level to a user, it will automatically be assigned to that user when he logs on to your application. During a session, a user can always override his current authorization level using the **File-Authorization-Level** menu (see also Section 20.3.4).

Rules   During user log on, AIMMS determines the authorization level applicable for that user in the following order:

- the authorization level of the user account itself,
- the authorization level of the first parent user group for which an authorization level has been specified, or
- the globally defined default authorization level (see also Section 20.3.2).

Password   In the **Password** tab of the **Properties** dialog box of either a user or user group you can specify a password for that user or user group. If you check the **Password Required** check box, an end-user is not allowed to 'enter' an empty password through the **File-Authorization-Change User Password** menu. The user password is verified whenever a user logs on to your application. The user group password is verified when a user wants to change his user group during a session via the **File-Authorization-Group** menu.

Developer rights   To every user or user group you can assign developer rights. Only users with developer rights are allowed to open a project, which is linked to a user database, in developer mode. If no user has been assigned developer rights, you can still open the project in developer mode by using the predefined "*User Administrator*" account.

ODBC/OLE DB user name and password   In the **Database** tab of the **Properties** dialog box of users and user groups you can enter a user name and password which AIMMS will use for authenticating the ODBC/OLE DB connections of a user. The following rules apply.

- If an ODBC/OLE DB user name and password have been specified for a user account itself, AIMMS will use these when authenticating an ODBC/OLE DB connection.
- Otherwise, AIMMS will inherit the ODBC/OLE DB user name and password of the first user group in which the user account is contained that

provides an ODBC/OLE DB user name and password, and makes these available to all of its children (as indicated by the **Inherit** check box).

- If no ODBC/OLE DB user name and password are found in the previous steps, or when ODBC/OLE DB authentication fails, a log on dialog box will be presented.

Within the modeling language AIMMS provides access to the currently logged-on user and his user group. They are available through:

Use within model

- the string parameter CurrentUser holding the name of the user currently logged on, and
- the string parameter CurrentGroup holding the name of the currently active user group.

By default, AIMMS does not provide access to the entire set of users and user groups defined in the user database attached to a project, as this information is not necessary for most applications. However, if you need access to the set of all users and groups in your application, AIMMS offers the following two functions to obtain this information.

Obtaining all users and groups

- SecurityGetUsers(*user-set*[,*group*][,*level*])
- SecurityGetGroups(*group-set*)

The argument *group* is a string, the argument *level* an element of AllAuthorizationLevels, while the output arguments *user-set* and *group-set* are (root) set identifiers.

The functions SecurityGetUsers and SecurityGetGroups create new elements in the indicated sets for every user or group in the user database. When specified in a call to SecurityGetUsers the group and level arguments serve as filters, filling *user-set* with only those user names that are member of the given group and/or possess the given authorization level.

Calling semantics

20.3.4 Logging on to an AIMMS application

Whenever an AIMMS application has an associated user database, you must first log on before you can run the application. The **Logon** dialog box is illustrated in Figure 20.9. Initially, AIMMS will enter your Windows user name if this

Logging on

Figure 20.9: The **Logon** dialog box

name is also present in the user database. You can always close an application during the log on procedure by pressing the **Cancel** button, if you do not have a valid user account for the application. After logging on successfully, AIMMS will set the user group and authorization level to the values associated with the account of the currently logged on user.

Switching authorization or group

Through the **File-Authorization** menu end-users can log off, or modify their current user group or authorization level if this is needed to read or write particular case data, or when additional authorization is required to perform particular tasks within the model. If you do not remove the (standard) **File-Authorization** menu from your application, you are strongly advised to password-protect all user groups and/or authorization levels to prevent unauthorized access by end-users.

Changing end-user passwords

Through the **File-Authorization-Change User Password** menu end-users of your application can modify their password without needing the interaction of the user administrator. By checking the **Password Required** check box in the **Properties** dialog box of an end-user, you can prevent end-users from entering empty passwords.

20.4 Case file security

Protecting your data

When your AIMMS-based application is used by multiple end-users all sharing the same data management tree, read and/or write protection of the individual datasets and cases may become a relevant issue. AIMMS offers such protection by associating cases and datasets with end-users in the user database.

Access rights

As explained in Section 20.3, user groups in the user database can be ordered in a hierarchical fashion. All case file security in AIMMS is based on this hierarchy. More specifically, AIMMS allows you to assign different access rights to

- the owner of the dataset or case,
- members of the group associated with the dataset or case,
- members of groups that lie hierarchically above the user group associated with the dataset or case,
- members of groups that lie hierarchically below the user group associated with the dataset or case, and
- all other users.

For each category of users you can separately specify read and write access to the case or dataset.

Only when you are the owner of a dataset or case, or are the local user administrator, will AIMMS allow you to modify the access rights previously assigned to a case. You can perform this task through the **Properties** dialog box of the dataset or case in the data manager. In the **Access rights** tab of this dialog box, which is displayed in Figure 20.10, you can change the associated user group, as well as the access rights for the each of the categories listed above.

Modifying access rights

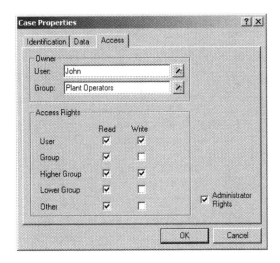

Figure 20.10: Access rights of a dataset or case

Normally, AIMMS will only allow you to modify the access rights of the datasets and cases you own. You can override this by checking the **Administrator Rights** check box displayed in Figure 20.10. This will open a password dialog box requesting the user administrator password associated with the end-user database. If successful, you can modify the access rights of any dataset or case as if you were its owner. With user administrator rights, you can even change the owner and user group associated with the case or dataset.

Administrator rights

By default, any newly created dataset or case will be owned by the user that is currently logged on, and will be associated with the currently active user group (usually the group in which the end-user is placed in the end-user database). The access rights associated with such a dataset or case will be the default access rights assigned to the end-user in the end-user database by the local user administrator.

Default access rights

You can specify the default access rights of an individual user or of an entire user group through the **Access** tab in the properties dialog box of either the user or user group at hand. In this dialog box, illustrated in Figure 20.11, you can either

Specifying default access rights

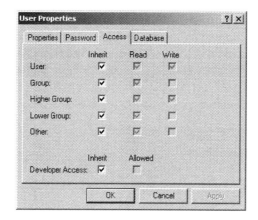

Figure 20.11: Specifying default access rights

- specify the specific access rights for a particular user category in a similar fashion as for a case or dataset itself, or
- indicate that you want to inherit the rights for a particular user category from the next higher user group.

Chapter 21

Project Settings and Options

Several aspects of AIMMS, including its startup behavior, its appearance, the inner workings of the AIMMS execution engine or the solvers used in a session, can be customized to meet the requirements of your project. This chapter describes the various tools available in AIMMS for making such customizations.

This chapter

21.1 AIMMS execution options

Many aspects of the way in which AIMMS behaves during a session can be customized through the AIMMS execution *options*. Such options can be set either globally through the options dialog box, or from within the model using the OPTION statement. As every project has its own requirements regarding AIMMS' behavior, option settings are stored per project in the project file.

Options

AIMMS offers options for several aspects of its behavior. Globally, the AIMMS execution options can be categorized as follows.

Option types

- **Project options:** how does AIMMS behave during startup, and how does AIMMS appear during a project.
- **Execution options:** how does the AIMMS execution engine with respect to numeric tolerances, reporting, case management and various other execution aspects.
- **General solver options:** how does AIMMS behave during the matrix generation process, and which information is listed.
- **Specific solver options:** how are the specific solvers configured that are used in the project.

Through the **Settings-Project Options** menu you can open the global AIMMS **Options** dialog box illustrated in Figure 21.1. In this dialog box, an *option tree* lists all available AIMMS execution and solver options in a hierarchical fashion.

Option dialog box

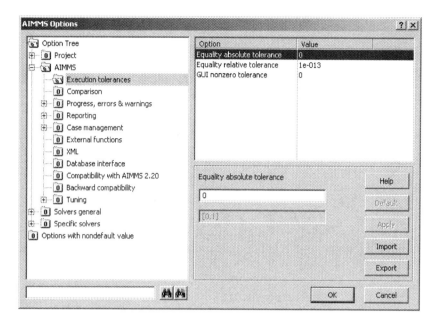

Figure 21.1: The AIMMS **Options** dialog box

Modifying options

After selecting an option category from the left-hand side of the **Options** dialog box, you can modify the values of the options in that category on the right-hand side of the dialog box. As illustrated in Figure 21.1, AIMMS lists the currently selected value for every option (in the first edit field) along with the allowable range of all possible option values (in the second field). Option values can be either integer numbers, floating point numbers or strings, and, depending on the option, you can modify its value through

- a simple edit field,
- radio buttons,
- a drop-down list, or
- a wizard in the case where the value of an option is model-related.

Committing options

With the **Apply** button, you can commit the changes you have made to the value of a particular option and continue changing other options; the **OK** button will commit the changes and close the option dialog box. With the **Default** button at the right-hand side of the dialog box, you can always reset the option to its default value. It is only active when the option has a nondefault value.

Option description

When you have selected an option, and need to know more about its precise meaning before changing its value, you can press the **Help** button at the right-hand side of the options dialog box. As illustrated in Figure 21.2, this will open a help window containing a more detailed description of the selected option.

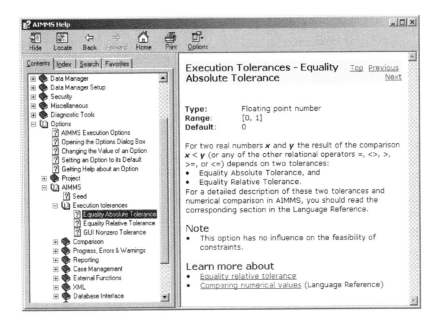

Figure 21.2: Option help

To help you quickly identify all the options which you have modified for a particular project, all modified options are summarized at the end of the options tree in a special section, **Options with nondefault value**. You can modify these options either in this section, or in their original locations. If you set a modified option back to its default value, it will be removed from the nondefault section.

Options with nondefault value

When you know (part of) the name of an option, but do not know where it is located in the option tree, you can use the search facility in the lower left-hand part of the option dialog box to help you find it. When you enter (part of) an option name, AIMMS will jump to the first option in the tree whose name contains the entered string.

Searching for options

In addition to modifying option values in the options dialog box, you can also set options from within your model using the OPTION statement. The OPTION statement is discussed in the AIMMS Language Reference. While changes to option values in the options dialog box are stored in the project file and reused at the beginning of the next project session, run time option settings are lost when you close the project. Setting options during run time can be convenient, however, if different parts of your model need different option settings.

Setting options within the model

21.2   End-user project setup

Setting up an end-user project

A number of options and settings are of particular importance when you want to set up a project in such a manner that it is ready to be used by end-users. You can find these options in the **Project-Startup & authorization** and the **Project-Appearance** sections of the **Options** dialog box. This section discusses the most important options.

Startup mode

Through the option *startup mode* you can specify whether your project should be started in developer mode or in end-user mode by default. Access to developer mode may be prohibited by a VAR license, or protected by a developer password (see Section 20.2). In addition, the default startup mode may be overruled either by command line options, or by holding down the **Shift** key during start up.

Startup procedure

With the *startup procedure* option you can select a procedure within your model which you want to be executed during the start up of your project. Such a procedure can perform, for instance, all the necessary data initialization for a proper initial display of the end-user GUI automatically, thus preventing your end-users from having to perform such an initialization step themselves.

Startup page

With the *startup page* option, you can indicate the page which AIMMS will display at start up. It is important to specify a startup page for end-user projects, as all data communication with the model must take place through end- user pages designed by you. Therefore, you should also ensure that every relevant part of your application can be reached through the startup page.

Project title

By default, AIMMS will display the name of the currently loaded project in the title bar of the AIMMS window. Using the *project title* option you can modify this title, for instance to provide a longer description of your project.

21.3   Solver configuration

Configuring solvers

With every AIMMS system you can obtain a license to use particular solvers to solve mathematical programs of a specific type. As AIMMS provides a standardized interface to its solvers, it is even possible for you to link your own solver to AIMMS. This section provides an overview of how to add solvers to your system or modify the existing solver configuration.

You can obtain a list of solvers currently known to your AIMMS system through the **Settings-Solver Configuration** menu. This will open the **Solver Configuration** dialog box illustrated in Figure 21.3. The dialog box shows an incidence

Solver configuration dialog box

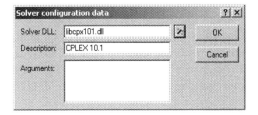

Description	LP	MIP	NLP	QP	MIQP	QCP	MIQCP	MCP	MPCC	MINLP	Solver DLL
AOA					x		x			**X**	libaoa.dll
BARON 7.5.3		x	x	x	x	x	x			x	libbaron753.dll
CONOPT 3.14A	x		**X**	x		x					libconopt314A.dll
CPLEX 10.1	x	**X**		**X**	**X**	**X**	**X**				libcpx101.dll
CPLEX 9.1	**X**	x		x	x	x	x				libcpx91.dll
KNITRO 5.0	x		x	x		x		x	**X**		libknitro50.dll
PATH 4.6								**X**			libpath46.dll
SNOPT 6.1	x		x	x		x					libsnopt61.dll
XA 14	x	x		x							libxa14.dll

Figure 21.3: The **Solver Configuration** dialog box

matrix between all available solver and types of mathematical programs. An 'x' indicates the capability of a specific solver to solve mathematical programs of a particular type. A bold '**X**' indicates that the specific solver is used as the default solver for mathematical problems of a particular type.

The buttons on the right-hand side of the dialog box let you globally modify the solver configuration of your AIMMS system. Through these buttons you can perform tasks such as:

Modifying solver settings

- modify the default solver for a particular model type, and
- add or delete solvers.

With the **Set Default** button you can set the default solver for a particular type of mathematical program. AIMMS always uses the default solver when solving a mathematical program of a particular type. A run time error will occur, if you have not specified an appropriate solver.

Selecting default solver

When you want to add an additional solver to your system, you can select the **Add** button from the **Solver Configuration** dialog box, respectively. This will open a **Solver Configuration Data** dialog box as shown in Figure 21.4. In this

Adding a solver

Figure 21.4: The **Solver Configuration Data** dialog box

dialog box you have an overview of the interface DLL, the name by which the solver is known to AIMMS and any appropriate arguments that may be needed by the solver.

Select solver DLL

In the **Solver DLL** area of the **Solver Configuration Data** dialog box you can select the DLL which provides the interface to the solver that you want to link to AIMMS. AIMMS determines whether the DLL you selected is a valid solver DLL, and, if so, automatically adds the solver name stored in the DLL to the **Description** field.

Solver arguments

In the **Arguments** area of the **Solver Configuration Data** dialog box you can enter a string containing solver-specific arguments. You may need such arguments, for instance, when you have a special licensing arrangement with the supplier of the solver. For information about which arguments are accepted by specific solvers, please refer to the help file accompanying each solver.

Installation automatically adds

When you install a new solver from the AIMMS installation CD-ROM, AIMMS will automatically add the solver to the **Solver Configuration** dialog box. If the newly installed solver is the first solver of a particular type, AIMMS will also automatically make the solver the default solver for that type. Thus, after installing a new AIMMS system, you do not have to worry about configuring the solvers in most cases, provided of course that your AIMMS license permits the use of the solvers you have installed.

Using a nondefault solver

By modifying the value of the predefined element parameter CurrentSolver in the predefined AllSolvers during run time you can, at any time during the execution of your model, select a nondefault solver for a given mathematical programming type that you want AIMMS to use during the next SOLVE statement for a mathematical program of that type. At startup, AIMMS will set CurrentLPSolver to the default LP solver as selected in the solver configuration dialog box.

21.4 Print configuration

Print configuration

AIMMS offers two distinct facilities to create printed reports associated with your model, namely printouts of graphical end-user pages and print pages (see Chapter 14), and printouts of text files such as an ASCII representation of a part of the model tree or the listing, log and PUT files. This section explains how you can configure the printing properties for both types of reports.

Printing end-user pages

End-user pages and print pages are printed according to the settings that you have selected for these pages. These settings include:

- the selection of the paper type on which pages are printed (see Section 14.1), and

■ the selection of object fonts and colors through the AIMMS font and color selection dialog boxes (see Section 11.2).

These settings must be fixed by you as the application developer, and cannot be changed by an end-user of your application. An end-user can, however, still select the printer to which the output must be sent, as explained below.

Text files can be printed from within AIMMS, either from the **File-Print** menu inside an AIMMS text editor window, or through a call to the FilePrint procedure from within a procedure in your model. The print properties of all text files that you want to print, in either manner, can be modified through the **Settings-Text Printing** menu. This will invoke the dialog box illustrated in Figure 21.5.

Text printing

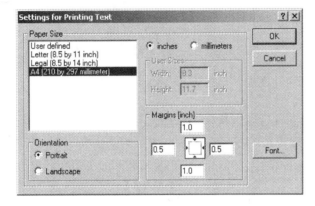

Figure 21.5: The **Text Printing** dialog box

In the **Text Printing** dialog box you can select the paper type and font with which you want all text files to be printed. For the paper type you can select one of the predefined paper types, or specify a user defined paper type by providing the page height and width, as well as the margins on each side of the page. By pressing the **Font** button on the right-hand side of the dialog box, you can select the font with which you want your text files to be printed. The text printing properties are stored globally on your machine.

Text printing properties

With the **File-Print Setup** menu you can select the printer on which print pages and text files associated with your project are printed, and modify the properties of that printer. This command will invoke the standard Windows **Print Setup** dialog box illustrated in Figure 21.6.

Printer setup

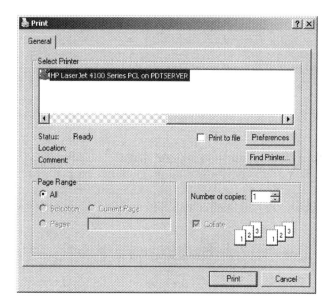

Figure 21.6: The **Print Setup** dialog box

Default settings   The settings selected in this dialog box will only be valid during the current session of AIMMS. If you want to modify the default print setup globally, you can do this through the **Printer** section in the Windows **Control Panel**. There you can

- select a **Default** printer from the list of all printers available on your system, and
- modify the **Document Defaults** (i.e. the printer settings with which each print job is printed by default) for every individual printer on your system.

Without a call to the **File-Print Setup** dialog box, AIMMS will use the default printer selected here, and print according to the document defaults of that printer.

Chapter 22

Localization and Unicode Support

When you are creating an end-user interface around your modeling application, you will most likely create the end-user interface in either your native language or in a common language like English. Which language you choose most probably depends on the intended user group of your application. In the case that you are requested to distribute your application to end-users who are not fluent in the language in which you originally developed the end-user interface, AIMMS offers a localization procedure which automatically separates all static texts used in the end-user interface of your application. This allows you to provide a relatively smooth translation path of your application to the native language(s) of your end-users.

Interface localization

If you have end-users on the Asian market who require a native version of your AIMMS application, only making use of AIMMS' built-in localization procedure is not sufficient, as Asian languages require the use of double-byte characters to represent native strings. To support you in such cases, AIMMS is also available in a separate Unicode version. The AIMMS Unicode version allows you to use (double-byte) Unicode characters in strings and set element descriptions in both your model and its end-user interface, and offers full support for communicating with files and databases containing either ASCII or Unicode data.

AIMMS *Unicode version*

This chapter illustrates how to use the automated localization procedure built into AIMMS, and explains how you can use it to create a foreign version of an end-user application. In addition, it describes the capabilities and limitations of the AIMMS Unicode version, as well as the necessary steps to run a project with the AIMMS Unicode version.

This chapter

22.1 Localization of end-user interfaces

Conceptually, localization of an end-user application consists of a number of basic steps. These basic steps are to

Basic concepts

- find all the strings that are used in the pages and menus of your end-user interface of your application,
- store these strings separate from the other interface components, and

■ provide translations in different languages of these separately stored strings.

Through the **Tools-Localization** menu, AIMMS offers an integrated localization tool which can perform the first two steps for you automatically. The result is a list of strings, each with a description of its origin, which can be easily translated to other languages. This section will explain the use of the localization tool built into AIMMS step by step.

Setting up localization support

Before you can start the final localization conversion of your AIMMS application, AIMMS needs to

■ add a **Localization** section to your model which contains a default setup for working with a localized end-user interface, and
■ register the names of the identifiers and procedures which are necessary for storing, loading and saving the strings used in the end-user interface of your application.

You can perform these steps through the **Tools-Localization-Setup** menu. As a result, AIMMS will add the (default) **Localization** section to your model if such a section has not already been added before. Secondly, through the dialog box presented in Figure 22.1, AIMMS will request the names of the identifiers

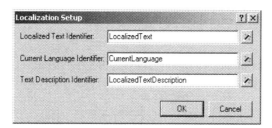

Figure 22.1: Setting up localization support

to be used further on in the localization process to store the strings used in your end-user interface. By default, AIMMS proposes the identifiers added for this purpose to the (newly added) **Localization** section. If you change the names of these identifiers, or want to use completely different identifiers, you can execute the **Tools- Localization-Setup** menu again to specify the modified names.

Localization section

After the localization setup has been executed for the first time, your model has been extended with a new section called **Localization**. The contents of this model section is illustrated in Figure 22.2. The declaration section contained in it declares the default set and string parameters used for storing all localization information.

■ The set AllLanguages contains the names of all languages to which you want to localize your application. You can add as many languages to

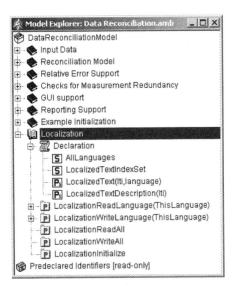

Figure 22.2: Localization section in the model tree

its definition as necessary. *However, you should make sure that, at any time, the first element in the set is your development language*: during the conversion process described below, AIMMS will associate all strings in the end-user interface with the first language from the set AllLanguages.

- Associated with the set AllLanguages is an element parameter CurrentLanguage, through which you (or your end-users) can select the language in which all texts in the end-user interface are to be displayed.

- The set LocalizedTextIndexSet is a subset of the predefined set Integers, and is used to number all strings within your end-user interface that are replaced by AIMMS during the conversion process.

- The string parameter LocalizedText contains the actual texts for all string objects in your end-user interface for one or more languages. During the localization conversion process, AIMMS will fill this parameter with the texts of your development language.

- The string parameter LocalizedTextDescription contains a short description of the origin of all converted string objects, and is filled by AIMMS during the localization conversion.

Through the **Tools-Localization-Setup** menu, you can modify the localization parameters which AIMMS will use during any subsequent conversion process. If you choose to select different identifiers, you should make sure that:

Using other localization identifiers

- the identifier selected for the **Localized Text Identifier** is a 2-dimensional string parameter, the identifier selected for the **Current Language Identifier** is a scalar element parameter, and the identifier selected for the **Text Description Identifier** is a 1-dimensional string parameter.

- the second index set of the **Localized Text Identifier** and the range set of the **Current Language Identifier** coincide. AIMMS will interpret the resulting set as the set of all languages.

- the first index set of the **Localized Text Identifier** and the first index set of the **Text Description Identifier** coincide and is a subset of the predefined set Integers. AIMMS will use this set to number all string objects during the conversion process.

Localization procedures

In addition to the sets and string parameters discussed above, the **Localization** section also contains a number of procedures added for your convenience to perform tasks such as:

- loading and saving the localized text for a single language,
- loading and saving the localized texts for all languages, and
- to initialize support for a localized end-user interface.

The statements within these procedures refer to the default localization identifiers created by AIMMS. If you have chosen different identifiers, or want to store the localization data in a nondefault manner, you can modify the contents of these procedures at your will. You must be aware, however, that the facilities within AIMMS to view and modify the localized text entries do not use these procedures, and will, therefore, always use the default storage scheme for localized data (explained later in this section).

The initialization procedure

The localization procedure **LocalizationInitialize** added to the **Localization** section of your model will read the localized text for a single language. If the element parameter CurrentLanguage has been set before the call to Localization-Initialize, AIMMS will read the localized strings for the language selected through CurrentLanguage. If CurrentLanguage has no value, the procedure will read the localized strings for the first language (i.e. your development language).

Added to Main-Initialization

If your model contains the (default) procedure MainInitialization (see also Section 4.2), a call to the procedure **LocalizationInitialize** will be added to the end of the body of MainInitialization during the first call to the **Tools-Localization-Setup** menu. This makes sure that the localized strings on pages and in end-user menus of a converted end-user interface contain the proper (original or localized) texts when the project is opened.

Performing the localization conversion

Through the **Tools-Localization-Convert** menu you can instruct AIMMS to replace all static string occurrences in your (end-user and print) pages, templates and end-user menus by references to the localization identifiers selected during the localization setup. During the conversion, AIMMS

- scans all pages, templates and menus for static strings,
- creates a new localized entry in the **Localized Text Identifier** for each such string, and

■ in the interface component where the static string was found, replaces it by the corresponding reference to the **Localized Text Identifier**.

In addition, AIMMS will, for each localized string, create a description in the **Localized Text Description Identifier**, initialized with the name of the page or menu plus the object in which the corresponding string was found. This may help you to link localization texts to specific objects and pages.

String description

During the localization conversion, AIMMS will warn for any duplicate string it encounters. For such duplicate strings, you have the opportunity to create a new entry in the **Localized Text Identifier** or to re-use an existing entry. Re-using existing entries can be convenient for common strings such as "Open" or "Close" that occur on many pages.

Duplicate occurrences

Once you have performed the localization conversion, you can view all localized strings through the **Tools-Localization-Show Strings** menu, which will open the dialog box illustrated in Figure 22.3. In this dialog box, AIMMS dis-

Editing localized strings

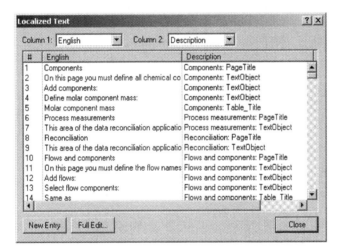

Figure 22.3: The **Localized Text** dialog box

plays a numbered list of all localized strings, along with the description of the origin of each string. The string numbers exactly correspond to the elements of the set LocalizedTextIndexSet discussed above.

Through the drop down lists at the top of the **Localized Text** dialog box of Figure 22.3, you can select the contents of the first and second string columns, respectively. For each column, you can select whether to display the localized text for any language defined in the set AllLanguages, or the description associated with each string. By viewing the localized strings for two languages alongside, you can easily provide the translation of all localized strings for a

Modifying dialog box contents

new language on the basis of the localized strings of, for example, your development language.

Modifying multiline strings   If a localized string consists of multiple lines, you can invoke a multiline editor dialog box to edit that string through the **Full Edit** button at the bottom of the **Localized Text** dialog box, as illustrate Figure 22.4. To invoke this multi-

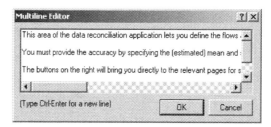

Figure 22.4: The **Multineline Editor** dialog box

line editor for the string corresponding to a particular language, click on the localized text for that language, and press the **Full Edit** button. The multiline editor will now be opened with the exact string that you selected in the **Localized Text** dialog box.

Localizing new texts   If you have added new pages, page objects, or end-user menus to your project after running the localization conversion procedure for the first time, you have two options to localize such new interface components. More specifically, you can

- localize every new component separately through the **Localized Text** wizard present at all text properties of the object, or
- run the localization conversion procedure again.

The **Localized Text** *wizard*   Whenever a string is associated with a property of a page, page object or menu item, the wizard button █ of such a property in the **Properties** dialog box provides access to the **Localized Text** wizard, as illustrated in Figure 22.5 Invoking

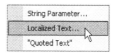

Figure 22.5: The **Localized Text** wizard

this wizard will open the **Localized Text** dialog box illustrated in Figure 22.3, in which you can either select an existing localized string, or create a new entry through the **New Entry** button. After closing the dialog box, AIMMS will add a reference to the localized text identifier in the edit field of the property for which you invoked the wizard, corresponding to the particular string selected in the **Localized Text** dialog box.

If you have added several new interface components without worrying about localization aspects, your safest option is to simply run the localization conversion procedure again. As a result, AIMMS will re-scan all pages, templates and menus for strings that are not yet localized, and add such strings to the list of already localized texts as stored in the localization identifiers associated with your project. Obviously, you still have to manually provide the proper translations to all available languages for all newly added strings.

Performing the conversion procedure again

By default, AIMMS stores the localization data as *project user files* containing standard AIMMS data statements within the project file (see also Section 2.5.2). The localized strings for every language, as well as the string descriptions are stored in separate user project files, as illustrated in Figure 22.6. The read

Localized text storage

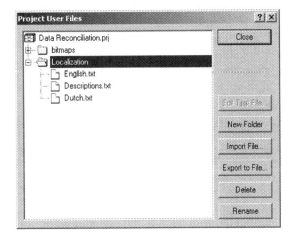

Figure 22.6: Default of localization data as user project files

and write statements in the bodies of the localization procedures added to the **Localization** section of your model, assume this structure of project user files for localization support.

Whenever you use the **Localized Text** dialog box of Figure 22.3, either through the **Tools-Localization-Show Strings** menu or by invoking the **Localized Text** wizard, AIMMS will make sure that the contents of appropriate localization data files are read in before displaying the localization data for a particular language. Likewise, AIMMS will make sure that the contents of the appropriate project user files are updated when you close the **Localized Text** dialog box.

Automatically updated

By using the import and export facilities for project user files (see also Section 2.5.2), you can also edit the data files containing the localized strings outside of AIMMS. This can be a convenient option if you hire an external translator to provide the localized texts for a particular language, who has no access to an AIMMS system. Obviously, you have to make sure that you do not

Manual edits

make changes to these files through the **Localized Text** dialog box, while they are exported. In that case, importing that file again will undo any additions or changes made to the current contents of the project user file.

Static strings in the model

Besides the static strings in the end-user interface of your AIMMS application, the model itself may also contain references to static strings or to sets whose elements are defined within the model itself. Such strings and set elements are left untouched by AIMMS' localization procedure. If your model contains such string or set element references, you still have the task to replace them by references to a number of appropriate localized string and element parameters.

22.2 The AIMMS Unicode version

AIMMS Unicode version

When you need to distribute a localized version of an AIMMS application to, for instance, Asian or Russian end-users, use of the common single-byte AIMMS version may not be sufficient anymore, as many languages in these regions cannot be represented by means of single-byte characters. To support localization to such languages, a Unicode version of AIMMS is available, in which all strings are represented internally through double-byte characters.

Installing the Unicode version

The AIMMS Unicode version is distributed as a separate installation program, and is available on the AIMMS installation CD-ROM or from the AIMMS website. Running the AIMMS Unicode installation program will install the AIMMS Unicode version alongside the ordinary single-byte AIMMS version. The AIMMS Unicode version can be run with your existing AIMMS license. When a valid single-byte AIMMS version has already been installed on your computer, the Unicode installation procedure will automatically copy your existing license files to the AIMMS Unicode installation directory.

Increased memory requirements

When you use the AIMMS Unicode version, the set element descriptions and the data of string parameters, as well as any other string data used in the model or end-user interface will consume twice as much memory as with the single-byte AIMMS version. In most cases, this will result in a moderate increase in memory usage. If your model contains a lot of set elements or string data, however, you may want to make sure the memory usage of the AIMMS Unicode version is still acceptable.

Converting to the Unicode version

When you have developed an AIMMS application using the ordinary (single-byte) AIMMS version, you can relatively easily convert your project to the AIMMS Unicode version, as all single-byte strings can be represented by double-byte Unicode strings without problems. As of yet, however, the binary .amb model files used by the single-byte and Unicode AIMMS version are incompatible. The project files used by both version are compatible. The next section explains how you can prepare your projects for use with the Unicode version.

To prepare your project for use with the AIMMS Unicode version, the following steps are required:

Conversion plan

- open your project with the single-byte AIMMS version, and open your model with the **Model Explorer**,
- save your model as an ASCII .aim file through the **File-Save As** menu,
- associate the newly created .aim file with your project through the **File-Open-Model** menu (thereby selecting the .aim file),
- save the project, and re-open the project with the AIMMS Unicode version.

After these steps you can proceed developing your project with the Unicode version. If you make changes your model, the model will be saved again in a binary .amb file. This .amb file will, however, contain the model text in Unicode format, and is incompatible with the single-byte AIMMS version.

By saving your model in an .aim file in the AIMMS Unicode version, you can also convert your project back to the ordinary single-byte AIMMS version. To create an ASCII .aim file in the AIMMS Unicode version, which can be read in by the single-byte AIMMS version, you must make sure that the option aim_output_character_set is set to ascii prior to saving the model to an .aim file. *Note, that for the back-conversion to be successful, you must make sure that your Unicode project or model file does not contain any genuine Unicode characters (i.e. double-byte characters which are not representable as single-byte characters).* In that case, the corresponding texts cannot be read into the ordinary single-byte AIMMS version, and loss of parts of your model or end-user interface may result.

Converting to the single-byte version

Both the single-byte and the Unicode AIMMS versions have been extended with Unicode I/O capabilities. The following list contains the Unicode I/O capabilities of both versions.

I/O capabilities of the Unicode and single-byte versions

- Both the single-byte and Unicode AIMMS versions can read and write to database tables containing either ANSII or Unicode string fields. The single-byte AIMMS version will only accept Unicode string data, as long as the ODBC or OLE DB driver manager is able to convert the Unicode string to a single-byte ASCII string.
- The internal text editor built into the AIMMS Unicode version will accept both Unicode and single-byte ASCII text files. Upon saving, AIMMS will only save the file as a Unicode file if it actually contains non-ASCII characters. The text editor of the single-byte AIMMS version only accepts ASCII text files.
- The DEVICE attribute of the FILE declaration (see also Section 27.1 of the Language Reference) has been extended with the additional devices disk(ASCII) and disk(Unicode) besides the existing disk device. The following rules apply for PUT, DISPLAY and WRITE statements that refer to the FILE identifier in the AIMMS Unicode version:

       – if the device is disk(ASCII), AIMMS will always create an ASCII out-
       put file, and complain if some of the output contains non-ASCII
       characters,

       – if the device is disk(Unicode), AIMMS will always create a Unicode
       output file, regardless whether the output actually contains non-
       ASCII characters,

       – if the device is disk, AIMMS will create an ASCII output file if the
       option default_output_character_set assumes the value ascii and a
       Unicode output file if the option assumes the value unicode. By de-
       fault, the option assumes the value automatic, in which case ASCII
       output will be created with the single-byte AIMMS version and Uni-
       code output with the AIMMS Unicode version.

The single-byte AIMMS version will always create ASCII output files, regardless of the settings of the DEVICE attribute.

- The READ statement in the AIMMS Unicode version will accept both ASCII
and Unicode data files. The READ statement in the single-byte AIMMS
version only accepts ASCII data files.

- A WRITE statement in the AIMMS Unicode version to a data file which is
not indicated through a FILE identifier, will create an ASCII or Unicode
output file depending on the option default_output_character_set. The
WRITE statement in the single-byte AIMMS version always creates an ASCII
data file.

- The listing file is created by the AIMMS Unicode version as either an ASCII
or a Unicode file depending on the option listing_file_character_set. A
Unicode listing file may be necessary, for instance, to display a constraint
listing referring to Unicode set elements.

The Unicode-related options can be found in the **AIMMS-Reporting-Unicode
support** folder of the **Options** dialog box (see Section 21.1).

*Arguments of
external
procedure calls*

When your AIMMS project makes use of functionality provided by functions
in external DLLs linked to your project, you can specify whether string arguments are to be passed as ASCII or Unicode character buffers. For every STRING
PARAMETER argument of an EXTERNAL PROCEDURE or FUNCTION, you can specify the
ascii and unicode properties in the PROPERTY attribute. The following rules
apply.

- If you specify the ascii property, both the single-byte and Unicode AIMMS
versions will pass the string argument as an ASCII string buffer (or an
array of string buffers for a multi-dimensional string parameter). An
runtime error will result, if the string data contains genuine Unicode
characters.

- If you specify the unicode property, both the single-byte and Unicode
AIMMS versions will pass the string argument as an Unicode string buffer
(or an array of such buffers).

Any string argument in a function of the AIMMS API, expects a Unicode character buffer in the AIMMS Unicode version, and an ASCII character buffer in the single-byte AIMMS version. Also, the data type `AimmsValue` and `AimmsString` expect either Unicode or ASCII string data, depending on the AIMMS version. In both cases, the `.Length` field of the data type refers to the length of the string in the appropriate character set, rather than the byte length of the supplied buffers. Therefore, if your executables and DLLs make use of the AIMMS API, you should make sure that the types of all string arguments passed to AIMMS through the API are of the appropriate type, depending on the AIMMS DLL version you are running your AIMMS project with.

Arguments of the AIMMS *API*

.

Appendices

Index